Handbook of common methods in
LIMNOLOGY

OWEN T. LIND

Department of Biology and
Institute of Environmental Studies,
Baylor University,
Waco, Texas

SECOND EDITION

with 40 illustrations

KENDALL/HUNT PUBLISHING COMPANY
4050 Westmark Drive   Dubuque, Iowa 52002

To
Robert S. (Bob) Campbell
teacher, gentleman, and friend

Copyright © 1985 by Kendall/Hunt Publishing Company

Copyright © 1979 by The C. V. Mosby Company

Previous edition copyrighted 1974

Library of Congress Catalog Card Number: 85–81316

ISBN 0–8403–3753–1

Printed in the United States of America
10   9   8   7

☐ Acknowledgments

Thanks are due the following individuals, whose comments, recommendations, and criticisms have helped me make this new edition more responsive to the needs of the students and faculty who use the book:

J. Foster III, *Tennessee Technological University*
J. Reynolds, *University of Victoria (Canada)*
A. Baker, *University of New Hampshire*
W. Widmer, *University of Connecticut*
J. Reynolds, *University of Missouri at Columbia*
S. Datt, *Loyola Campus of Concordia University*
R. Bailey, *University of London*
J. Olive, *University of Akron*
C. Falter, *University of Idaho*
W. Ratzlaff, *Millersville State College*
L. Oglesby, *Pomona College—Claremont*
R. Morris, *Laurentian University (Canada)*
G. Hennings, *Kean College of New Jersey*
L. Pomeroy, *University of Georgia*
L. Williams, *University of Alabama*
W. O'Brein, *University of Kansas*
L. Jahn, *Western Illinois University*
K. Wood, *State University College (Fredonia, N.Y.)*

☐ Contents

4   Benthos, 153

Appendices

Introduction

The study of limnology is, in its broadest sense, aquatic ecology. The biological limnologist is primarily concerned with the interactions between aquatic organisms and their physical and chemical environments. Because of the more sharply defined environmental limits and further advanced methods of analysis, limnologists are often able to quantify environmental factors more precisely than are their terrestrial counterparts.

Biologists are occasionally reluctant to attempt essential physical and chemical water analyses because of lack of confidence or even boredom with nonliving systems. This reluctance produces a superficial treatment of ecosystem components or, worse, erroneous physical and chemical data. However, the critical interdependencies of aquatic life and the physical and chemical environment, along with the investigator's awareness of the dynamics of such a system, make total ecosystem analysis not only essential, but fascinating as well.

This handbook is designed for those who wish a collection of established methods in a form convenient for use in either the field or laboratory. Although written primarily from the freshwater viewpoint, most of the procedures may be used directly or with slight modification in marine and estuarine ecosystems. No chemical procedures requiring more than simple instruments such as the Spectronic 20 or pH meter are included. The Ap-

1

pendices, which provide commonly used conversions and equivalents and sources of equipment and supplies, will be especially valuable to the new student or occasional investigator.

The techniques chosen for inclusion are primarily descriptive. Although modern-day limnology as a science is concerned with experimental studies of the dynamics of aquatic ecosystems and their biota, beginning students must first grasp certain basic descriptive tools. These tools enable the students to know the "setting" on which the process being experimentally investigated is played. Correct determinations of many chemical and biological parameters are essential to the experimental process. It is probable that many of the aquatic science students of today will have little opportunity for the excitement of experimental discovery but will be trained for and pursue careers involving routine survey and descriptive tasks. Because of the tendency of the beginner to accept anything in print as fact, it is essential (1) that the methods included are proven reliable through years of use over a variety of water types, but (2) that the beginner approach each technique in a flexible and critical manner.

Because of rapidly changing instrumentation technology and limnological theory, no completely up-to-date methods book is possible. *Limnological Methods* (Welch, 1948), although dated, contains much useful information, especially on morphometry, plankton, and benthos. It is of limited value in other areas, such as chemistry and biotic dynamics. Fortunately, more recent books on chemical analysis are available. *Methods for Collection and Analysis of Water Samples* (Rainwater and Thatcher, 1960) and, more recently, *Methods for Collection and Analysis of Water Samples for Dissolved Minerals and Gases* (Brown, Skougstad, and Fishman, 1970) are methods manuals used by the Chemical Quality of Water Branch, Geological Survey, U.S. Department of the Interior. The first sections of these manuals are devoted to excellent discussions of collection and analytical techniques as well as instrumentation evaluation.

A preferred source of chemical limnological methods is *Standard Methods for the Examination of Water and Wastewater* (American Public Health Association, 1976; hereafter referred to as *Standard Methods*). This is a compilation of methods agreed on as standard by three major groups involved in water analysis: the American Public Health Association, the American Water Works Association, and the Water Pollution Control Federation. Wherever applicable, analysts are encouraged to use the tested methods of this book and to report findings in the prescribed terms. This has the advantage of placing the data of many investigators on a comparable basis and allows for more valid comparisons across the country. Often the investigator may simply cite *Standard Methods* in his publications and, because of the wide availability of this source, no further description of analytical methods is required.

A Practical Handbook of Sea Water Analysis (Strickland and Parsons, 1968) is a widely acclaimed reference book for limnologists as well as those dealing with marine ecosystems. *Methods for Chemical Analysis of Water and Wastes* (Environmental Protection Agency, 1974) is a compilation of methods by the Analytical Quality Control Laboratory of the Environmental Protection Agency for use in the agency's laboratories. It describes methods that are either those given in *Standard Methods* or modifications thereof as required by the agency. It also includes techniques to permit automated chemical analysis for many water constituents.

The Hach Chemical Co., Ames, Iowa, has been successful in manufacturing prepackaged reagents for chemical methods of analysis that do not assume any chemical understanding on the part of the investigator. Many of their prepackaged chemical methods are taken from *Standard Methods* and work well.

No single method will be adequate for all waters. Because of unusual concentration ranges of the element of interest or the presence of unusual and unsuspected interfering substances, the beginning investigator should be especially wary of accepting any analytical results until he has confirmed that the method of choice, whether prepackaged chemicals or detailed laboratory technique, provides him with correct data. Confirmation of results can be obtained by comparing data with those published from the same or similar waters or by cross-referencing against an alternative method.

Standard Methods also contains sections on bacteria, plankton, and benthos; the latter two discussions are essentially the same as in Welch's text (1948). However, since most aquatic biologists tend to develop and prefer their own techniques for collection and analysis of biological specimens, there are few methods that can truly be considered standard. A welcome addition for the biological limnologist was the English translation of *Methods of Hydrobiology (Fresh-water Biology)* by Schwoerbel (1970). This book is especially valuable to beginning aquatic biologists because of the comprehensive subject-listed literature references. Several of the International Biological Program handbooks will also be of interest to freshwater biologists: *A Manual on Methods for Measuring Primary Production in Aquatic Environments* (Vollenweider, 1969), *Methods for Chemical Analysis of Fresh Waters* (Golterman and Clymo, 1969), *Methods for Assessment of Fish Production in Fresh Waters* (Ricker, 1968), and *A Manual on Methods for the Assessment of Secondary Productivity in Fresh Waters* (Edmondson and Winberg, 1971).

There are several small, taxonomically oriented works on freshwater biology, such as those by Needham and Needham (1962) and Eddy and Hodson (1961). The most comprehensive work on freshwater biology is *Fresh-water Biology* (Edmondson, 1959), an excellent taxonomic source, especially for invertebrates. Another is *Fresh-water Invertebrates of the United States* by Pennak (1978). For identification of algae, the books by Smith (1950) and Prescott (1962, 1970) will be helpful. Patrick and Reimer's work (1966) is the best for information on diatoms. Fassett (1960) identifies aquatic flowering plants, and Eyles and Robertson (1963) and Correll and Correll (1972) identify plants of southern waters.

Obviously, serious students of limnology with professional ambitions will want to purchase most of these books for their own libraries; however, for all but the most fortunate, this acquisition must be a gradual process. It is for beginning students that this handbook has been prepared. It is designed to provide adequate information for the beginner but is by no means intended to replace any of the works cited previously. Most of the methods contained herein are not mine; they represent a collection of methods that have been found to provide relatively accurate results with a minimum of complicated procedures. Most are seasoned with advice and comments derived from several years of experience with them in introductory limnology classes at Baylor University and the University of Michigan Biological Station.

1 □ Physical limnology

MAPPING
Mapping a pond or small lake by plane table method

The mapping of lakes is usually beyond the capability of the limnologist, who must rely on maps made by well-equipped survey teams. Often aerial photographs with known scale will suffice. However, small ponds (generally less than 5 hectares) may be accurately mapped with a minimum of equipment. Whenever possible, mapping in winter on sound ice is preferable. Welch (1948) includes an extensive section on map methods from simple to complex.

Apparatus

Plane table and tripod
Alidade (crude alidade may be made by setting pins in opposite ends of wooden ruler for ''sights'')
Compass
Ruler, graduated in millimeters
Map paper
Hard-lead pencils
Round-headed map pins
Steel measuring tape, approximately 30 to 50 m long
2 steel stakes, approximately 0.5 m long
Wooden stakes, approximately 1 m long (25 to 100, depending on size of pond)
Sledgehammer
Stadia rod or other painted pole at least 2 m long

Base line

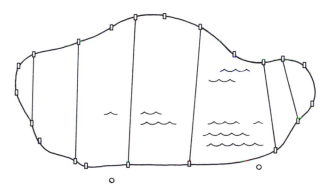

Fig. 1. Plane table map *(top)* showing position of numbered shoreline stakes and the base line. Distances for three shoreline points as determined by triangulation from the ends of the base line are shown. In a small lake or pond, bottom contours may be determined by sounding along a calibrated rope stretched between approximately opposite shoreline stakes *(bottom)*.

Metal-core clothesline rope, of sufficient length to reach across
    pond, marked in 1 and 5 m increments
Small boat or canoe
Depth sounding line or electronic depth sounder

Procedure

1. Select the longest relatively straight section of
shoreline for establishing a base line.

2. Drive steel stake near shore for one end of base
line. Most of pond should be visible from this point. Repeat for other end of base line at a point at least 30 m from
first stake and approximately the same distance from
shore as the first. Make the base line as long as possible
(Fig. 1).

3. Drive a wooden stake at water's edge at every
major change in shoreline configuration. Maximum distance between any two wooden stakes should probably
not exceed 20 m. Drive wooden stakes just deep enough
to stay in place for mapping. Steel stakes on base line are
permanent and should be driven so only 2 to 5 cm pro-
trude. Tie a cloth "flag" to top of every fifth stake to facilitate counting.

4. Attach map paper to plane table. Set up tripod directly over one metal stake at end of base line, and plumb
table center to the stake. Level table, and align one edge
with true north-south line. Lock table in place. Draw true
north arrow in corner of map; also indicate magnetic
north (Fig. 2).

5. Judge shape of pond and length of base line, and
determine appropriate scale. Record scale on map near
north-south arrow.

6. Determine position on map that represents end of
base line where table is set up. Mark this on map by
firmly setting map pin in table. The pin must not move.
Check again to be certain that all parts of pond will fall on
map according to scale selected and position of pin.

7. Have flagman set stadia rod vertically on steel rod
at other end of base line. Place zero mark on alidade

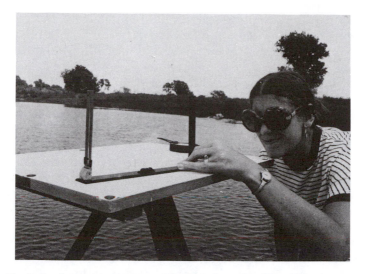

Fig. 2. Student using an open-sight alidade and a plane table to map a small pond. A compass is used to ascertain north orientation of the map. The edge of the alidade at the end nearest the student is always placed against a map pin firmly set in the table. This pin is the map position corresponding to the base line stake over which the table is centered.

against pin, and sight through alidade down base line to line up with stadia rod (Fig. 2). Draw base line on map along edge of alidade for the appropriate length as determined by scale. *Be careful not to bump the plane table at any time.*

8. Now in a similar manner, progressively sight alidade around the pond on each wooden stake (flagman holds stadia rod just behind the stake), and draw line along edge of alidade. Number each line consecutively. The fifth line and every multiple of 5 should be sighted on a stake with flag. Any deviation indicates a missed stake. After each flagged wooden stake has been sighted and the line drawn, resight on other end of base line. If this sighting does not coincide with original line, the table has been moved and another series must be taken.

9. After a complete circuit has been made, move and reset table over stake at other end of base line. Set pin in map at this end of base line and align table by sighting back down to other end of base line. Repeat step 8.

10. The point at which corresponding numerical lines intersect is the position of each wooden stake marking the shoreline. Connect these points with a smooth line. By doing this in the field, you are able to include minor changes in shoreline. Also indicate position of obvious features in the water (logs, weed beds, and\others).

11. It is more difficult to map bottom contours for the purpose of morphometric calculations. The problem is the accurate plotting on the map of the position of the sounding team. On smaller ponds, mapping bottom contours may be done as follows: starting at one end of the pond, stretch the calibrated rope between any two known wooden stake locations. Represent this by a light line on the map.

12a. The sounding team in the boat moves along this line, taking soundings at appropriate intervals from one shore to the other (intervals vary depending on amount of change in bottom contours). With the use of the scale, a point is placed at the proper position on the line, and the measured depth written in at this point (Fig. 1).

12b. An alternate procedure, more suitable for larger bodies of water but less precise so far as exact position on the lake is concerned, is the use of timed echo soundings. Echo sounders are becoming increasingly available, especially as "fish finders." The boat's pilot makes a line of sight between two stakes on opposite banks and sets the boat at a constant low speed. It is important that he make every effort to maintain a straight line and constant speed between the two points. A second person calls out regular time intervals—for example, every 30 seconds—and a third person reads and records the depth shown on the echo sounder at that time. The total time taken to transect the lake is recorded. Thus the total time for distance is known, and time for any timed increment can be calculated. Although this procedure is less precise, it has a compensating advantage in that it allows many more soundings to be taken, which more accurately gives the shape of the lake basin and allows for the plotting of more closely spaced bottom contours. This increases the accuracy of the lake volume calculation.

13. Repeat the chosen sounding procedure for a parallel series of lines across the pond.

14. Determine degree of bottom development to be shown (1 m intervals are often used, but intervals may be more or less). Draw in contour lines by connecting the appropriate points and the lines. For shape between points, parallel the shoreline.

15. Label map and include date and names of map crew. Roll up map (do not fold).

16. Return map to laboratory; retrace shoreline and contours in permanent ink. Clean up other working lines with soft eraser.

Direct measure modification of plane table method

The direct measure modification of the plane table method is suited for ponds of regular outline and open water. It is simple, and depth soundings may be taken simultaneously if the work party is of sufficient size.

Apparatus

Same as for plane table method plus the following:

Good quality ruler; meter stick cut off at 60 to 70 cm is convenient length

Steel-core clothesline rope, of sufficient length to reach across pond's longest dimension, marked in 1 and 5 m increments.

If approximate size of pond is known before going to field, previously prepared table of scale conversions of meters to millimeters will facilitate time spent in field

A small boat or canoe may be necessary if obstructions are present

Procedure

1. Select a base point that is relatively open of vegetation and from which all parts of the pond may be seen.

Drive iron stake and center drawing board over this stake. All work will be done from this location.

2. Prepare board for drawing as in plane table procedure.

3. Set out shoreline stakes as in plane table procedure.

4. Select point on paper to correspond with plane table setup over iron stake, and set map pin.

5. Use alidade and take sightings on first stake to left. Draw light line along alidade from base point out to near edge of paper.

6. Place zero end of calibrated line on iron stake, and measure distance to the sighted stake. Convert this distance to scale in millimeters; using ruler, measure out from base point the corresponding distance on the line,

Fig. 3. Compensating polar planimeter used to determine areas. Here the planimeter is tracing the shoreline to determine the total area of a pond drawn by the plane table method.

and mark point. Thus location and distance of first stake have been determined.

7. Repeat step 6 for all remaining stakes.

8. Connect the points, filling in detail from observation.

9. If the map party is of sufficient size, a boat sounding team may work alongside the measuring rope, taking soundings and distance between base point and stake. Depths called out may be marked on map at proper distance from base point (see following section on morphometry).

MORPHOMETRY

Most limnological phenomena, distribution of biota, and productivity are directly related to the morphological features of the water basin. Therefore certain morphometric features are of interest to a limnologist beginning a study of any water. Morphometric measurements are based on good hydrographic maps, and, in general, the larger the maps, the more reliable the morphometric data that may be obtained from them.

Area by polar planimeter

Whenever instrumentation is available, the polar planimeter method is preferred. The polar planimeter (Fig. 3) is a delicate instrument, and care must be taken whenever it is used. Read over the directions supplied by the manufacturer for the instrument in use. Be sure to observe instructions regarding direction of tracing (usually clockwise) and minimum and maximum allowable angles of the tracing arm.

1. Prepare map by placing it on a hard, smooth surface. Tape it in place.

2. Most maps will be too large to be covered by one cycle of the planimeter. These maps must be ruled off into halves or smaller segments, and the areas of each summed for the total.

3. Check the calibration of your instrument on the

map paper being used. Use calibration device supplied by manufacturer, or carefully rule off a known area and trace with planimeter three times.

4. Proceed with actual planimetry by tracing outline of lake or bottom contours, following carefully the manufacturer's directions.

Area by cut and weight

1. Lightly trace map outline with bottom contours onto a good grade of paper.

2. From an area outside traced area, cut out a square of known area (a 10 cm × 10 cm square is usable) and weigh this piece of paper. Calculate weight per square centimeter.

3. Cut out outline of entire lake and weigh. Calculate area by dividing weight of entire lake by weight of 1 square centimeter.

4. Repeat step 3 for each successive bottom contour.

Volume by calculation

If a lake basin is considered as a cone, then the volume may be calculated by the appropriate equation (cone volume = $1.047 \, r^2 h$). However, because the slopes of lake bottoms are rarely regular, a better approximation for volume may be obtained by calculating and then summing the volumes of conical segments (frustra), with upper and lower surfaces delimited by the areas of sequential depth contours. The calculation is then as given by Welch (1948):

$$\text{Lake volume} = \sum \text{Frustrum volumes}$$

where

$$\text{Frustrum volume} = \frac{h}{3} \, (a_1 + a_2 + \sqrt{a_1 a_2})$$

h = Depth of frustrum
a_1 = Area of frustrum surface
a_2 = Area of frustrum bottom

Volume by planimetry

Lake volume may also be determined by planimetric integration.

1. Using linear graph paper, plot the area at a given depth against that depth (hypsographic curve). Make the plot with the horizontal axis for area at the top and the vertical axis for depth at the left of the page. This places the 0—0 ordinate in the upper left corner.

2. Integrate the area beneath the curve by using a polar planimeter or by counting squares. Include squares that are more than half within, and disregard those which are more than half outside the line. If equal dimensions for each square of the grid are used (that is, each grid has dimensions of 1 m^2 in one direction versus 1 m in the other direction), that grid represents 1 m^3 volume. Summing the number of cubic meter volume grids under the curve will give the approximate total volume of the lake. NOTE: This is the same form of integration described on p. 135.

Shoreline length: cartometer method

The map measure (cartometer) is a convenient method for measuring lengths of shorelines (Fig. 4). It is also a delicate instrument and deserves care in handling.

1. Set dial by turning wheel to zero line. Draw a line of known length on the map paper, and trace three times with the instrument to check its accuracy of calibration.

2. Set instrument to zero line, and carefully trace the shoreline of the lake. Watch carefully to see if the dial revolves more than one time. Record number of inches or centimeters, and convert to feet or meters per scale.

3. Repeat for each of the submerged contours.

Shoreline development

Shoreline development (SLD) is an index of the regularity of the shoreline. For a lake that is a perfect circle, the shoreline development is 1. As the value departs from unity, irregularity is indicated. A few natural lakes,

Fig. 4. Cartometer in use measuring the shoreline length of a plane table map.

such as craters and sinkholes, are near unity. Elongate lakes and especially reservoirs impounding water in valleys and side valleys may have SLDs of more than 4. This calculation is of limnological interest because it relates to age—that is, older lakes have had more shoreline erosion, thus are more regular in outline—and to total productivity through more or less littoral areas. This value is calculated as follows:

$$\text{SLD} = \frac{S}{2\sqrt{a\pi}}$$

where

S = Length of shoreline
a = Area of lake

Maximum length and orientation of main axis

These two factors are usually the same portion of the lake, but because of unusual irregularities, they may not be so. Maximum length is the longest straight line that may be drawn without intersecting any mainland. Not all lakes will have a segment that may be considered to represent a maximum length. The orientation is expressed as opposing points on a 16-point compass, for example, SSE–NNW. The orientation may be of great significance because it relates to the prevailing seasonal winds of the region.

Maximum depth

The maximum depth (Z_{max}) is the deepest spot in the lake.

Mean depth

The mean depth $(\overline{Z})$ is an important value. It is used in certain other calculations, such as heat budgets, and is generally considered as a reliable morphometric indicator of trophic conditions with $\overline{Z}$ inversely related to production. It is calculated as follows:

$$\text{Mean depth in meters} = \frac{\text{Volume in m}^3}{\text{Surface area in m}^2}$$

STREAM FLOW

Physical measurements of flowing waters not only include temperature and light, as in lakes, but also rate of flow and volume of discharge. Stream flow rate is highly variable in different stream portions such as rapids, riffles, and pools. However, it is also variable within one of these types of stream segments. In straight-flowing segments, friction along the stream bottom and sides produces "drag" with maximum velocity occurring in midstream near the surface. In stream bends, the maximum velocity is obtained near the surface at the outside of the curve. Biologists should note that measurement of velocity in such open water portions has little meaning in describing the habitat of most organisms, who usually seek some association with the stream bottom or obstructions to flow. Such considerations are well described by Hynes (1970). Volume of discharge is often significant to the limnologist who wishes to determine the downstream transport of dissolved and suspended matter, stream biota, or both.

Velocity

Stream velocity is best measured at several portions of the stream by a flowmeter. However, in the absence of such a mechanical device, the rate of movement of a neutrally buoyant object will provide a reasonable estimate.

Apparatus

Neutrally buoyant object: sphere such as an orange (the fruit) works well (polyethylene bottle filled with water also works well, although not as easily seen as the orange; the main concern is to choose some object that is visible but does not project above stream surface where it will be affected by wind)

Watch (stopwatch, preferably)

Tape measure, 25 to 50 m

Procedure

1. Measure a length of stream, and mark start and finish. The longest reasonable length without changing stream condition is desired. The longer the segment, the more accurate the timing of the floating object.

2. Place the neutrally buoyant object at the starting point, and measure the time required for it to reach the finish point.

3. Repeat this at least three times, and average the measured times.

4. Express velocity as meters per second. If the measurements were made at midstream, thus at maximum velocity, the mean stream velocity is approximately 0.8 of the measured velocity (Welch, 1948).

Volume discharge

The discharge of a stream is a function of the mean velocity, as estimated above, and the cross-sectional area of the stream. The most accurate measurements of discharge are based on flow through a weir, which is a trough with sloping sides of exactly known cross-sectional area for a measured depth. However, a reasonable estimate of cross section may be made from depth soundings of the stream made in transect at the point at which velocity measurements were made. If a long stream segment was used for measurement of velocity, or if the stream bed has significant changes in configuration, several cross-sectional areas may be required and then averaged, since the velocity measured is also an "average" of the segment.

Apparatus

Sounding line: calibrated pole may be easier to use in shallow streams, whereas sonar depth sounder is aid for deep rivers
Tape measure
Linear graph paper

Procedure

1. Take a series of soundings along a transect across the stream.

2. Measure the length of the transect across the surface of the stream.

3. Using linear graph paper, draw the cross section of the stream to scale. Connecting the depths with a smooth curve is usually preferable to straight lines, which will tend to underestimate the area. Integrate the enclosed area by planimeter or by counting squares (see p. 135 for further discussion of integrations).

4. Calculate the volume discharge as follows:

Discharge (m³ per second) = Mean velocity (m per second) ×
Mean cross-sectional area (m²)

TURBIDITY

Turbidity in water is the presence of suspended solids, which reduce the transmission of light either through scattering or absorption. Athough the loss of light (extinction) is readily determined, the quantification relative to some standard for the suspended material is difficult because of its nonhomogeneous nature. However, a polymer, formazin, has come into wide use as a standard. Readings using a formazin suspension approximate those obtained by the earilier turbidity method, the Jackson Candle.

The use of a nephelometer, which measures the light scattered at a 90-degree angle, is preferred; however, a spectrophotometer may be used to measure the transmitted light, which is approximately inversely related to the amount scattered.

Nephelometry
Apparatus

> Nephelometer (Hach model 2100 or equivalent) with standard formazin series and spotless nephelometer tubes
> Graduated cylinder (25 ml)
> Kimwipes or other soft tissues

Procedure

1. Allow nephelometer to warm up according to manufacturer's instructions.

2. Set to full-scale concentration using the appropriate formazin standard (note the necessity of changing the height of standard above light source for certain concentrations). It is good practice to check the formazin standards against each other because the formazin suspension will deteriorate.

3. Mix the water sample well, but avoid shaking so as not to incorporate bubbles, and place the appropriate volume (25 ± 1 ml in Hach model 2100) in the nephelometer tube. The sample should be at room temperature to prevent condensation on the sides of the nephelometer tube.

4. Wipe all moisture and fingerprints from sides and bottom of tube, then place it in the instrument and cover to exclude light.

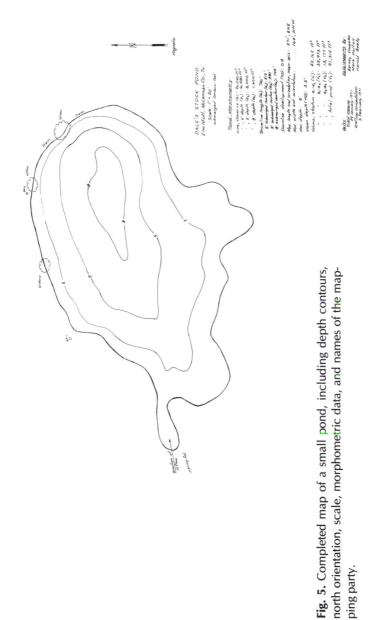

Fig. 5. Completed map of a small pond, including depth contours, north orientation, scale, morphometric data, and names of the mapping party.

Table 1. Jackson units (450 nm, 1-inch test tube)*

Meter readings	0	1	2	3	4	5	6	7	8	9
10	395	380	360	350	335	320	310	300	290	280
20	273	265	258	250	245	240	233	228	222	217
30	211	206	200	197	193	188	184	180	175	172
40	168	164	160	157	153	150	147	144	140	137
50	134	131	128	125	123	120	117	114	112	109
60	106	104	101	99	97	95	92	90	88	86
70	84	81	80	77	75	73	71	68	65	64
80	61	59	56	54	51	49	47	44	42	39
90	36	32	30	26	22	20	16	12	8	4

*From Hach Chemical Co., Ames, Iowa.

5. Read nephelometer turbidity units (NTU) directly from meter.

Spectrophotometer

Table 1, a table for turbidity, was made from standard formazin solutions, using a Jackson Candle Turbidimeter to measure the turbidity. The turbidity is expressed in standard Jackson units (JTU).

Apparatus

Spectronic 20 with blue-sensitive photocell set at 450 nm (see pp. 49 to 52 for instructions on use)
1-inch matched tubes

Procedure

1. Use demineralized water for blanking the Spectronic 20.

2. Mix room-temperature sample well by inversion; avoid shaking that would introduce bubbles.

3. Fill a 1-inch test tube with some of the turbid water, and read percent transmittance.

4. Find the turbidity units from the table. NOTE: Turbidity as JTU obtained by spectrophotometer is not directly comparable with data as NTU obtained by

nephelometer. However, for suspensions of clay less than 50 NTU, the relationship is *approximately* JTU = 2 NTU.

VERTICAL ILLUMINATION

Vertical illumination is defined as the illumination on the horizontal plane at some depth in the water, as measured by submarine photometer. Note that the photometer measures illumination and not energy flux or irradiance (which have dimensions of time and area, e.g., calories per square centimeter per minute). A photometer may be calibrated to read in footcandles or lux. Transmission of light energy through the water column is of special interest for many studies, especially those related to photosynthesis where one often wishes to determine a depth receiving 1% of the surface illumination. Several factors affect the transmission of light in water. Extinction of light results from one or more of three factors: the water itself, suspended particles, and dissolved material or color. The intensity of light penetration to depth z may be expressed as

$$I_z = I_o e^{-(n_w + n_p + n_c)z}$$

where

n_w, n_p, and n_c = Extinction due to each of the three factors, respectively

Poole and Atkins (1926) take the vertical extinction coefficient (n'') to be based on penetration of illumination disregarding the angular height of the sun. The validity of this assumption lies in the fact that as the sun becomes lower in the sky and fewer direct rays penetrate the water, a much greater proportion of sky radiation enters the lake with compensating effect. This can be shown empirically to be a valid assumption. This vertical extinction coefficient (n'') is a measure of the slope of the line that is obtained by plotting submarine photometer readings on the logarithmic axis against depth on semilogarithmic paper.

Pure water transmits light maximally at 460 nm in the blue region. Transmission in the red region at 680 nm is almost seventy times less. In natural waters the upper meter serves as an effective filter, removing much of the infrared and ultraviolet radiation. Transmission in the purest lakes approaches that of distilled water, with blue light penetrating to the greatest depths; this, plus the blue of the reflected sky, produces the apparent blue color of these clear lakes. With increasing turbidity, the peak of maximum transmission shifts toward the orange-red portion of the spectrum.

Submarine photometers, equipped with a selenium photovoltaic cell, have a response similar to that of the human eye but extending somewhat further into the blue region of the spectrum and slightly further into the red. The photovoltaic cell has a maximum sensitivity at 555 to 560 nm. It is important to keep in mind that this photocell is not equally efficient at all regions of the spectrum. For example, it is only approximately one fourth as efficient for light with a wavelength of 435 nm as it is at maximum sensitivity.

Some instruments include both a deck cell and a submersible cell. This allows intensity at any depth to be related to surface intensity at the same time. Having both types of cells is especially helpful when surface intensities are variable because of broken and moving cloud cover. Other instruments with only submersible cells adequately relate subsurface intensity to surface intensity by taking surface measurements at the beginning and at the end of the submerged series.

Procedure for single-cell instruments

Follow specific instructions of photometer manufacturer.

1. Open meter and check to see that controls are set at OFF and at the lowest amplification.

2. Adjust photocell so that it hangs horizontally from the suspending cables. If neutral density or colored

filters are used, make sure that all surfaces of glass are wet and free from bubbles.

3. Lower photocell into the water so that its surface is just wet. Record surface reading.

4. Lower photocell into the water to a second pre-scribed depth—usually 0.5 m will suffice—and take a second reading. Continue this procedure to a depth at which no further light is detected.

5. Return photocell to the surface in the same incre-ments, and retake readings on the way up. Record these values. Any large discrepancy from those obtained while lowering should be rechecked. Most investigators choose to take the average between the two values for the final light reading. Exercise caution when lifting the cell from the water to be sure that it is in a position at which the light will not be of sufficient intensity to cause the meter to go off scale; it is usually best to set the function switch to the off position when handling the photocell in the boat.

6. Repeat the entire procedure, substituting colored filters and removing neutral-density filters as required. If possible, use red-, green-, and blue-colored filters such as Schott RG-2, VG-9, and BG-12, respectively.

Plotting extinction curve

In plotting extinction curves one plots depth on the vertical, or Y, axis. Surface, or zero, depth is taken as the top of the figure, and depths are plotted downward as they naturally occur. With linear paper, a plot of pho-tometer readings against depth typically results in a smooth curve. An alternate plot producing a nearly straight line results when semilogarithmic graph paper (3 or 4 cycle) is used. Plot meter readings against depth, placing the meter readings on the logarithmic axis (Fig. 6). For the sake of comparisons, it is often preferable to plot percent of surface intensity versus the depth at which they occur (i.e., 100%, 75%, 50%, 25%, 5%, 1%) instead of absolute measured intensities.

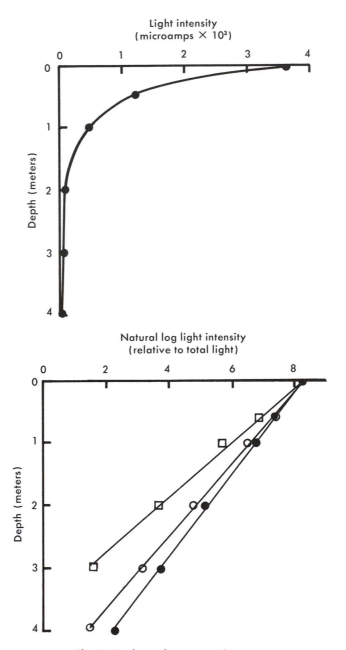

Fig. 6. For legend see opposite page.

Determining vertical extinction coefficient

The vertical extinction coefficient (n″), which describes the slope of the line determined before in the depth-versus-intensity plots, may be calculated by the use of the relation

$$I_z = I_o e^{-n''z}$$

where

I_z = Light intensity at depth z
I_o = Light intensity at surface
$n'' = n_w + n_p + n_c$

The least square estimate of n″ is given by:

$$n'' = \frac{\ln I_o \, (\Sigma z) - \Sigma [z \, (\ln I)]}{\Sigma z^2}$$

To solve this equation, divide a sheet of paper into five columns. Column one, to be titled z, is the column in which you will list all depths at which light readings were taken, starting with zero as the surface; column two titled I_z, is the column in which you will list the corresponding intensitites at each depth; column three, titled $\ln I_z$, is the column in which you will list the corresponding natural log for each intensity at depth z; column four, to be titled z ($\ln I_z$), is the column in which you will list the factor obtained for column three multiplied by column one; the fifth column, to be titled z^2, is the column in which you will list the square of each depth. Columns one, four, and five are to be summed.

Fig. 6. Light intensity versus depth plots. The rapid attenuation of illumination by water is readily seen from the top plot. However, for purposes of comparison, the use of semilogarithmic paper or the plotting of natural logarithms of the intensities as shown in the bottom plot is preferred. In this example, the differential transmission of red *(open circles)* and of blue *(open squares)* light as compared with the total illumination *(closed circles)* is apparent. The absolute surface intensity of red and of blue light was adjusted to provide overlying data points for the surface.

Fig. 7. The Secchi disk may be easily made from a 20 cm diameter metal or weighted wooden disk. Opposite quarters are painted gloss white and gloss black as shown. It is important that the calibrated line be attached so that the disk hangs horizontally in the water. (From Vivian, V. E.: Sourcebook for environmental education, St. Louis, 1973, The C. V. Mosby Co.)

The extinction of light between any two depths may be calculated more simply by the following equation:

$$n'' = \frac{\ln I_{z_1} - \ln I_{z_2}}{z_2}$$

This type of calculation allows for the detection of abnormally high or abnormally low light-altering strata within the total column of water when several of these values are compared.

VISIBILITY

Visibility is a measure of the depth to which one may see into the water. Obviously this is variable with the day conditions and the eyesight of the observer. The Secchi disk (Fig. 7) is a simple device used to estimate this depth. It consists of a weighted circular plate, 20 cm in diameter, with the surface painted with opposing black and white quarters. It is attached to a calibrated line by a ring at the center so that when held by the line, it hangs horizontally. To determine the Secchi disk visibility, slowly lower the disk into the water until it disappears, and note this depth. Lower the disk a little further, then slowly raise it until it reappears, and note this depth. The average of these two readings is taken for the final Secchi disk visibility depth.

The Secchi disk visibility is useful as a means of comparing the visibility of different waters, especially when measured by the same observer. Since clearness of the day, position of the sun, roughness of the water, and the observer all are significant considerations, they should be recorded along with the visibility depth data. Most important is for an observer to establish a standard set of operating conditions for himself; for example, always take readings while standing, with or without glasses or sunglasses, on the lee side of the boat with the sun to the observer's back, sometime between 9 AM and 3 PM.

Since one of the more frequently used optical relations in water studies is the photic depth, or depth of 1% surface illumination, some limnologists find it convenient to "calibrate their eyes" to estimate photic depth by using only the simple Secchi disk. This is done by accurately determining the true photic depth by use of a submarine photometer and at the same time taking a series of Secchi disk readings to obtain an average. Dividing true photic depth by Secchi disk visibility depth will obtain a factor by which Secchi disk visibility depth is multiplied. This factor is used in the future to estimate photic depth when a submarine photometer is unavailable. If this procedure is used over a range of different water conditions, one general factor may be developed and a fair approximation of photic depth obtained. The factor should fall between 2 and 5 for most natural waters without unusual turbid layers.

TEMPERATURE

Sufficient and accurate temperature data are important to the limnologist. Temperature directly and indirectly exerts many fundamental effects on limnological phenomena such as lake stability, gas solubility, and biotic metabolism.

Apparatus

Many satisfactory methods for determination of surface and subsurface temperatures are available, the main differences being in convenience of use. Most limnologists prefer to use the electric thermistor thermometer when one is available. However, a weighted, low-cost minimum-maximum thermometer provides an economical substitute. Because of the high specific heat of water, one may even bring samples to the surface and quickly measure temperature with a laboratory thermometer before the water mass has time to change temperature significantly. The brass-cased reversing thermometer was long a standard tool, but because of its relatively high cost and the lowering of prices of thermistor types, it is used much less today.

Procedure: thermistor thermometer

1. With the probe disconnected and the instrument turned on, set the mechanical adjustment of the meter movement, which is usually a screwdriver adjustment located on the case of the meter, so that the meter pointer is aligned with the scale mark. For maximum accuracy on all adjustments and readings, tap the meter *lightly* with the fingernail to overcome any friction in the meter movement, and place the instrument in its normal operating position.

2. Connect the probe, and take a shaded air reading of temperature. Then place the weighted end of the cable into the water at the surface, and hold it in place until the meter comes to a complete stop. This may require some time on the first reading since the metal weight tends to hold heat. Lower the cable and take readings at 1 m intervals unless the specific types of data required call for more frequent or less frequent readings.

3. For maximum accuracy, take a second series of readings while returning the probe to the surface, and take the average of the two measurements as the temperature for that depth.

4. Temperature data are frequently presented as a depth-temperature plot. Since the original definition of a thermocline was any depth at which temperature falls 1° C or more for each 1 m of depth, the usual plot uses equal units for temperature (° C) increasing from left to right on the upper abscissa and depth (m) increasing from zero at the upper left on the ordinate. Any slope equal to or greater than 45 degrees from the vertical may be defined as thermocline. In this context, refer to Hutchinson (1957) for a discussion of the metalimnion concept as a better interpretation of a depth-temperature profile.

Calculation of annual heat budget

Heat budgets are statements of the amount of heat energy entering or leaving a lake or pond. The one most commonly computed is the annual Birgean heat budget. This is a measure of heat energy input from the lowest heat content in winter to the greatest heat content in summer; thus it is the annual heat gain. It was first proposed by E. A. Birge and is represented by θ_{ba} with units of calories per square centimeter of surface. Heat budgets vary with both the size of the lake (as volume changes in relation to surface area) and with latitude (and to some extent altitude). A temperate lake at 40° N latitude will have a much greater temperature difference from minimum to maximum than will a tropical lake at 15° N latitude. Heat budgets reported for medium-sized to large ($\bar{z}$ greater than 10 to 15 m) temperate zone lakes tend to range from 10,000 to 60,000 calories/cm²/year. Heat budgets for smaller and shallower ($\bar{z}$ less than 10 m) temperate lakes and ponds and tropical lakes tend to be less than 10,000 calories/cm²/year. The heat budget may be calculated mathematically or determined from integration.

To calculate, the following equation may be used:

$$\theta_{ba} = \bar{z}(\bar{T}_s - \bar{T}_w)$$

where

$\bar{z}$ = Mean depth
$\bar{T}_s$ and $\bar{T}_w$ = Maximum summer mean and minimum winter mean temperatures

Where mean temperatures are not easily or reliably calculated, the heat budget may be determined by integration as follows:

1. Multiply the difference between the maximum summer temperature and the minimum winter temperature for each depth by the area (in centimeters) of the lake at that depth. NOTE: The area for each depth may be taken from the hypsographic curve of the lake (p. 15). If it is impossible to obtain winter minimum temperature data, one may assume the lake to have been homothermous at or near 4° C. However, this applies only to true temperate lakes; measurements must be made for tropical or subtropical lakes.

2. Plot the product of each of the preceding calculations against depth in centimeters, and connect the data points with a smooth curve.

3. Integrate the area under the curve by planimetry or counting the squares (see p. 135 for more on integration). Each square has dimensions of ($° C \times cm^2$) $\times$ cm. Thus an area is expressed in $° C \times cm^3$ or calories (1 calorie is the amount of heat energy necessary to change the temperature of 1 g [$1\ cm^3$] $1° C$).

4. Divide this total heat gain by the area of the lake's surface in centimeters to yield the heat budget as calories per square centimeter of surface.

2 □ Chemical limnology

COLLECTION OF WATER SAMPLES FOR CHEMICAL ANALYSIS

Since the mass of water being investigated is usually great relative to the sample, the sample is representative of the mass only to the extent that the mass is homogeneous. Such homogeneous conditions do not usually exist in the dimensions of either surface area or depth. Recent studies using aerial and satellite photography and imagery have made us especially aware of the nonuniform conditions that exist over the horizontal dimension of the lake surface. Thus care must be exercised in the selection of the location and number of sampling sites. Frequency of sampling varies with the intent of the investigation. For general information-gathering surveys, bimonthly series may suffice. Although many investigations are based on fewer samples, 24 can hardly be considered representative of the environmental conditions in a year.

WATER SAMPLERS

Although one may take water samples near the surface by simply filling a bottle or may dive to an approximate depth and open a bottle, a water-sampling device is desirable (Fig. 8). Such devices are designed not only to sample water from discrete depths but also to allow transfer of this water into the storage bottle without agitation or aeration. Several types of sampling bottles have been designed, including the Kemmerer, the Nansen, and the van Dorn. These are available in both metallic and plastic models. They differ principally in closing mechanisms. All are designed to cut a column of water so that when they reach the desired depth, they contain a sample of water from that depth. They are then tripped by a messenger and snap shut in various ways.

A sampler that is adequate for all collections except dissolved gases may be fabricated out of common materials as follows (Fig. 9). Obtain a glass bottle of $1/2$- to 1-gallon capacity. A bottle with a finger loop on the neck is ideal. Affix a weight to the bottom of this bottle sufficient to produce a negative buoyancy when the bottle is filled with air. This weight can be made by filling a coffee can with concrete. Bore a hole in a rubber stopper that fits the mouth of the bottle. Bolt an eyebolt through the stopper so that the eye is on top. Fasten the sampling line to the eyebolt approximately 50 cm from the end, and affix the end to the neck of the bottle or to the finger ring on the neck.

In use, the bottle is stoppered and lowered gently to the desired depth while suspended from the line attached to the eyebolt in the stopper. When the desired depth is reached, a sharp tug on the line will pop the stopper from the bottle, and the bottle will be suspended from the line secured to its neck. The bottle then fills at that depth.

A

Fig. 8. Three types of samplers used to collect subsurface water samples. **A,** The Kemmerer sampler has long been a favorite of limnologists because of its ruggedness and reliability. Although usually constructed of brass, it is now available in plastic (PVC) material for use where metal ions must be avoided.

B₁

Fig. 8, cont'd. B, The van Dorn sampler is usually made of PVC. It has the advantage, because of the external trip mechanism, of being attached to a line in series for winch and boom operation in deep waters. The van Dorn type is available in horizontal models, which permit sampling at discrete depths, especially useful in shallow water or near the lake bottom. A sampler rigged for horizontal use is shown in the "set" position (**B₁**). A transparent view (**B₂**) shows rubber tubing holding rubber closures in place. It is possible to fabricate a van Dorn sampler from PVC pipe, two rubber plunger (plumber's helpers) heads, a length of Latex tubing, and a trip mechanism.

Continued.

Fig. 8, cont'd. C, The Nansen bottle, easily attached in series to a line, is an established favorite of oceanographers and large-lake limnologists. Since closure is by rotation with the sampler inverted, Nansen bottles are frequently fitted with reversing thermometers so that exact temperatures at the depth of each sampling are measured. (Courtesy Wildlife Supply Company, Saginaw, Mich. 48602.)

Fig. 9. A simple water sampler of the Meyer type is easily constructed from a weighted bottle. This type of sampler is not suitable for collection of subsurface water for dissolved gas analysis.

An alternate procedure suitable for gas analyses that do not require extreme accuracy includes the use of the same type of weighted bottle suspended by a line about the neck. In this case a two-holed stopper is inserted in the neck of the bottle, with a small-bore glass tubing projecting through the stopper and reaching almost to the bottom of the bottle. In use, this bottle is lowered swiftly to the desired depth. Water flows in through the small-bore tubing; if the depth is not too great, little contamination with upper water occurs. The water flowing through the tubing into the bottom of the bottle displaces the air above it, producing a minimum amount of agitation and surface contact with the air. It must be emphasized, however, that this method is crude and allows for some gas contamination both at the air-water interface in the bottle and through the bubbling of the air out of the hole at the mouth.

SAMPLE STORAGE

Unless specifically stated otherwise, samples may be collected and then stored in Pyrex or polyethylene bottles, but soft glass bottles may suffice if analysis is to be performed within 48 hours. New bottles, as well as other new glassware, should be filled with water and allowed to soak for several days to leach out any water-soluble substances in the glass. Sample bottles must be cleaned after each use as described in the next section.

In the field, the sample bottle should be rinsed with the water to be sampled before it is filled. Certain factors, such as dissolved gases, are especially susceptible to change after sampling; for these, analysis must be done in the field, and this is so noted with the procedure. Most other factors are subject to varying degrees of change between time of collection and time of analysis. Biological activity within the sample is a major problem. It is recommended that samples be placed immediately in an ice chest, covered with crushed ice, and the lid replaced to exclude light. These may be transferred to a refrigerator in the laboratory while awaiting analysis. However, the temperature specified for the procedure must be noted, and time allowed for the sample to warm to this temperature.

If the analyses are to be performed within a few days, refrigeration at 4° to 5° C is often adequate. However, one must always consider the possibility of error introduced by bacterial activity, chemical precipitation, or adsorption onto the walls of the container. Various prestorage treatments or preservation may be desirable and may include filtration (separation into particulate and soluble fractions), acidification with HNO_3 or H_2SO_4 to a pH of about 2, or bacterial inhibition with mercuric chloride (40 mg/L). Be sure that the added material will not interfere with a possible analysis, such as mercuric chloride interference with the cadmium reduction method for nitrate nitrogen.

LABORATORY PROCEDURES
Cleaning glassware

Good technique must be emphasized. The key to this lies in the skill, experience, and, most important, the attitude of the technician. Many data have been discarded after a careful inspection of suspect data has indicated a hurried and careless technique or the lack of attention to detail. One of the best places to develop technical proficiency is in the care of analytical glassware. Many analyses are for trace amounts of an element in extremely diluted solutions (microchemistry on macroquantities). To prevent contamination, glassware must be spotless. A routine cleaning procedure is as follows:

1. Always empty sample water out as soon as you are sure you are finished with it; rinse bottle with tap water. This prevents the formation of silt and other scums on the bottom, which are difficult to remove.

2. Wash bottle with warm water and a special low-sudsing, phosphate-free glassware detergent. Use brushes when necessary, but avoid scratching with the

wire core because any roughening of the surface increases the probability of contamination.

3. Rinse three times with tap water.

4. Rinse with 5% HCl if analysis is for any element
likely to be found in detergent.

5. After the acid rinse, rinse three more times with
tap water.

6. Give a final rinse, inside and out, with distilled
water.

7. Dry and store in a dust-free place.

In the event that this procedure does not satisfactorily
clean the glassware, a chromic acid cleaning solution
may be used. This is prepared by adding 100 ml concentrated H_2SO_4 slowly and carefully to 3.5 ml saturated
sodium dichromate solution. *Handle this with caution.*

Grease may be removed with organic solvents such as
acetone (be careful of vapors: avoid prolonged breathing
and NO FLAMES) or soaking for about 15 minutes in a 2%
(w/v) warm NaOH solution. CAUTION: Excessive NaOH
may corrode glass surfaces; always rinse with diluted HCl.

Precautions

Before beginning any analytical procedure, always
read through the entire procedure. Such a practice will
save many samples that would otherwise be ruined and
lost. It will also save time, since many procedures call
for prior preparations or waiting periods during which
time other tasks may be done. Many of the reagents used
in these procedures are poisonous and require special
precautions. It does you no good to read that you are not
to pipette some poisonous substance by mouth after your
mouth is full of it. The many concentrated acid and alkaline solutions used in the limnology laboratory are
hard on both the skin and clothing; handle with care, and
always *wipe up spills immediately.*

Your determinations are only as good as your reagents; thus take special care to avoid contamination.

Keep the glassware covered with the proper stopper for that particular reagent. Do not use glass stoppers for alkaline solutions or rubber stoppers for organic solutions. Do not pipette directly from stock bottles; pour out the approximate amount needed into a clean beaker or flask, and pipette from it. Never return excess contents of such beakers and flasks or burettes and pipettes to the stock bottle. It is false economy to chance the ruin of a liter of reagent in order to save 10 ml.

Dilutions

When dilutions of reagents or standards are required, a simple relation for the proper proportion of the stock being diluted to diluent (usually demineralized water) is

Milliliters of stock to be diluted to desired volume =
$$\frac{\text{Desired concentration} \times \text{Desired volume (milliliters)}}{\text{Concentration of stock}}$$

where

Concentrations are normality, molarity, or weight or volume per volume

Occasionally crude dilutions, especially of acids, are given as parts of reagents plus parts of diluent on a basis of volume to volume, weight to volume, etc. For example, the notation 1 + 1 HCl (v/v) calls for equal parts of concentrated hydrochloric acid and demineralized water on a volume-to-volume basis.

Units used for reporting results of chemical analyses

Concentrations determined from analyses may be reported in several ways. The most commonly used form of reporting in freshwater studies is weight per volume, that is, milligrams per liter (mg/L) or micrograms per liter (μg/L), which will be the standard form used in this book. Other possible forms of reporting follow.

Parts per million (ppm) or parts per billion (ppb). These are often used interchangeably with mg/L or μg/L. This is valid for diluted solutions in which the specific gravity is near unity. However, this is incorrect for brines or severly polluted water such as mine drainages or seawater. For specific gravities much greater than unity:

$$ppm = \frac{mg/L}{\text{Specific gravity}}$$

Volume per volume (ml/L). This is occasionally used, especially in laboratory studies with manometry for dissolved gases.

Atomic weights per volume (mg-at/L). This is the form favored by oceanographers.

$$\text{Conversion: mg-at/L} = \frac{mg/L}{\text{Atomic weight of element}}$$

Molar concentration (mmoles/L). Occasionally used, especially for dissolved gases.

$$\text{Conversion: mg/L} = \text{mmoles/L} \times \text{molecular weight}$$

Equivalent weights per volume (mEq/L). This form is important because it permits direct summation of elements for mass balance determinations. It applies only to ionized substances. Convert weight/volume (mg/L) measurements to equivalent/volume (mEq/L) as follows:

$$mEq/L = \frac{mg/L}{\text{Sum of atomic weights/ionic charge}}$$

For example

$$25 \text{ mg } SO_4^{=}/L = ? \text{ mEq } SO_4^{=}/L$$
$$mEq \ SO_4^{=}/L = \frac{25}{\dfrac{96.062}{2}} = 0.52$$

DESIGN OF SAMPLING PROGRAM

As has been previously noted, a water sample from a lake or stream is representative of the true conditions present only so much as the water is homogeneous both through time and space. It is evident that such homogeneity is rarely if ever the case. Thus one may justifiably ask if the data obtained are valid. If the analytical procedure used is proved to be both accurate and reasonably precise, then one might expect any variation between two samples to be the result of true differences in the environment. How may we then use and express such variable data? Often the nature of the study will dictate the sampling regimen used. In some cases, complete randomization in time and all dimensions of space may be desirable, but more often, the data sought are based on some a priori knowledge of conditions. For example, if one were interested in determining the minimum dissolved oxygen content to which a particular community is exposed, one would find it necessary to sample for this parameter at some predawn time rather than at a more "convenient" midday time. Thus randomization in the time dimension is no longer possible. When one uses such previous knowledge, a stratified random sampling is used. That is, given that certain randomizations are precluded, one still attempts to randomize to the maximum extent possible.

If we then take an oxygen sample from a surface 1 m stratum at dawn, can we be assured that this sample truly represents the oxygen concentration to which the plankton community is exposed at that time? The obvious answer is that in the absolute sense, no, we cannot, but to gain further confidence we must take a second or replicate sample. Thus to be confident that our data truly represent the real conditions, our sampling regimen must have replication. Replication is very important in that it permits us to now consider some statistics about the population of data being sampled (population here is a statistical term not to be confused with the ecological term for a local assemblage of a species). The two types of statistical measurements of concern to most sampling programs are the measure of central tendencies and of dispersion. By far the most commonly used measure of a central tendency is the arithmetic mean ($\bar{x}$) or average. The mean is determined by dividing the sum of the individual observations (i.e., measurements) by the

number of observations or by using conventional notation, $\bar{x} = \Sigma x_i/n$. However, for most purposes, the mean
taken alone is inadequate and in some cases may be seri-
ously misleading. A statement of the degree of dispersion of individual sample measurements about the mean
is thus desirable. The commonly used measure of such
dispersion is the sample variance (s^2) or its more conve-
niently used square root, the standard deviation(s). Since
standard deviation is a statement of the dispersion of observations about the sample mean, the difference between each observation and the mean is the calculation
of interest. The standard deviation is then the square root
of the sum of the square of these differences divided by
the number of observations minus one, or, using conventional notation:

$$s = \sqrt{\frac{\Sigma(\bar{x} - x_i)^2}{n - 1}}$$

We often report our data as the mean plus or minus
one standard deviation, or using common notation, $\bar{x} \pm$
1s. Besides convenience, such reporting has interpreta-
tive value. If we assume that our samples were taken at
random and that all sample values are randomly distributed about the mean value, we may examine the area of
a normal curve encompassed by the mean plus or minus
one or more standard deviations. The mean plus or minus
one standard deviation will encompass about 66% of the
area, whereas two standard deviations encompass about
95%. In practice, then, this tells us that if we were to take
another sample, 95% of the time the measured value
would be within the range of values extending from two
standard deviations less than the mean to two standard
deviations greater than the mean. A second measure of
dispersion should be mentioned here and clearly distinguished from the standard deviation. This is the standard
error of the mean ($s_{\bar{x}}$). It is easily computed by dividing
the standard deviation by the square root of the number
of samples and is consequently always a smaller quan-

tity. However, this statistic describes the distribution of sample means about the mean. Thus it can provide an estimate of the probability of a new sample mean falling within certain limits about the original mean. One must always be careful to report whether a data report, that is, 6.5 ± 0.8, is the mean plus or minus the standard deviation or the standard error. Unless noted otherwise, that is, $\bar{x} \pm 2s$, it is assumed that the reported number is $1s$ or $s_{\bar{x}}$.

These statistics are the basis of many of the statistical tests employed in experimental limnology, such as testing for the differences among two or more means or data distributions. For the method of calculation of these statistics (many modern pocket calculators are now programmed for direct computation of mean and standard deviation) as well as the statistical tests, one should refer to one of the excellent statistic texts such as Steel and Torrie (1960), Sokal and Rohlf (1969), or Huntsberger and Billingsley (1977). Although obvious, it also should be emphasized that a mean and standard deviation drawn from multiple analyses of one large sample or aliquots from one water sample provides only an estimate of the precision of one's method and techniques. Variation in the environment must be determined by replicate individual samples.

TERMINOLOGY

Distilled water is water that has been obtained from an ordinary metal still. This water may contain ammonia and carbon dioxide gases, as well as traces of metals. It is used for routine dilutions in which the contaminations just cited will not be a problem and for the final rinse of glassware.

Glass-distilled water is distilled water that has been run through a Pyrex glass still. This process effectively removes most trace metals, but ammonia and carbon dioxide persist.

Demineralized water results from the passage of distilled water through a mixed-bed ion exchange column.

Fig. 10. Commonly used pipettes. The upper two are the serological type, and the lower two are the volumetric type. Both serological pipettes are calibrated to deliver (TD); the second is calibrated to the tip. To deliver 10 ml from this pipette, one must blow out the contents. This requirement is indicated by the encircling frosted glass ring on the mouthpiece *(arrow)*. The upper of the two volumetric pipettes is the most accurate of the four pipettes for delivering 10 ml. The calibration line *(arrow)* is in the small-diameter neck, which contains only a small percent of the total volume; thus any error in reading the meniscus is minimized. The contents of this pipette are not blown out, but drain when the pipette is in the vertical position with the tip against the side of the receiving vessel. A small amount of liquid must remain in the tip. The lower volumetric pipette is used to transfer volumes for which rapid dispensing is required. It has a large-bore tip, and the contents must be blown out (note frosted ring).

This produces high-quality water that is essentially free of ammonia but still contains carbon dioxide. This water is called for in the preparation of all reagents unless otherwise specified.

Carbon dioxide–free water is obtained by vigorously boiling demineralized water for a few minutes and cooling without agitation just before use.

Volumetric glassware may be calibrated either "to contain" (TC) or "to deliver" (TD). TD glassware will deliver only when the surface is absolutely clean so that no drops cling to the surface and when the tips of pipettes and burettes are absolutely smooth and free from chips. The contents of TD pipettes with a complete encircling frosted glass ring around the upper portion must be blown out for accurate delivery (Fig. 10). The contents of all other TD pipettes (the more common) are allowed to drain by gravity until a small amount remains in the tip; this tip is then touched to the side of the receiving vessel, and an additional small amount flows out by surface attraction. This type of pipette always retains a small amount in the tip that is not to be blown out. Volumetric pipettes and flasks are used when the procedure calls for a specified volume in milliliters, such as 1,000 ml, 250 ml, or 1.00 ml. Though not as accurate as volumetric pipettes, accurately calibrated serological pipettes may be used for measuring fractions of milliliters. When the volume is specified only as 1 liter or 10 ml, then other types of measurements may be used, such as graduated cylinders, for which a tolerance of ±5% is allowed.

In order that data may be uniformly interpreted by any investigator, it is necessary to use some standard as to how many decimals the results of analysis may be carried; in other words, the investigator must report only *significant figures.* The accepted procedure is to report the certain figures and only the first uncertain figure. For example, if a result of 11.17 ml/L is read from a curve but it is not known that the procedure is this accurate, the result should be reported as 11.1. If the investigator was certain of the 11 but was uncertain of the .1, the .1 could well be .0 or .2. If, on the other hand, experience has shown the procedure to be accurate and precise to an extent that 11.17 ml should have been reported, then the figure must not be rounded off to a small number as some may do in order that it fit into a column of figures. *The report should include figures justified by the accuracy of the work.* When figures are listed in a column, unwarranted zeros so that all will have the same number of places to the right of the decimal should not be included. However, zeros that are justified should not be dropped; if a burette is accurately read to 9.50 ml, the figure

Fig. 11. The easily used Spectronic 20 colorimeter-spectrophotometer. Note the mirrored meter scale. The meter is correctly read when no reflected meter needle image is seen. The small sample holder *(left)* may be replaced with holders accommodating sample tubes up to 1 inch in diameter, thus increasing sensitivity in quantitative analysis. Instrument controls are off-on and zero (dark-current) control *(lower left),* the 100% transmittance (light or gain) control *(lower right),* and the wavelength control *(right center).*

should not be rounded off to 9.5 ml. *Standard Methods* contains more discussion of this important subject.

Precision is a measure of the reproducibility of a method on a given sample under the same conditions. It is usually expressed as the standard deviation from the mean.

Accuracy is a statement of the error of the method and is usually expressed as a comparison of the amount of an element detected by analysis with the amount actually present. This is determined by analyzing for known standard quantities.

INTRODUCTION TO SPECTROPHOTOMETRY

One of the most important laboratory instruments in water chemical analysis is the spectrophotometer-colorimeter. The performance of the Spectronic 20 (Fig. 11) is between that of the simple colorimeter and the fine spectrophotometer. The ordinary colorimeter uses filters and therefore light from a large portion of the visible spectrum; a spectrophotometer separates light more exactly by a prism or grating. The best spectrophotometer uses nearly monochromatic light with a wave band width of less than 1 nm. The Spectronic 20 is a spectrophotometer whose band width is a constant 20 nm over the entire visible spectrum. Although this is far from monochromatic, it is adequate for many analyses and has become the workhorse of most laboratories.

The concentration of many substances may be determined by this instrument when a color is produced proportional to the concentration of the substance. The basis for this assumption rests in the Beer-Lambert law. The absorption of monochromatic light by a solute may be represented by the expression

$$I_z = I_0 e^{-bz}$$

where

I_z = Intensity of transmitted light
I_0 = Intensity of incident light

e = Base of the natural logarithms
b = Absorption coefficient due to the molecular extinction of light and the molar concentration of the solute
z = Thickness of the absorption cell in centimeters

In practical terms, this means that light passing through a given solute is absorbed in a constant exponential manner and that the total amount absorbed is a function of molar concentration, since other factors are constant for that solution. This relationship holds true in its most exact sense only for monochromatic light. When the Spectronic 20 is used, some deviation from a straight relationship must be expected. This difficulty is overcome by establishing a standard reference curve for each analytical method. At regular intervals, when a new reagent or set of reagents is made up, or whenever the results are suspect, a complete set of standards and a new calibration standard curve are prepared. The new standard curve consists of a *minimum* of five standards covering the range of concentrations expected in the water sample. It is wise to prepare one standard near the upper end of the curve each time an analysis is performed to check for deviation from the standard curve.

For each spectrophotometric procedure in this book, I have provided five suggested standard dilutions for preparation of standard curves with the concentration of the element being directly determined as milligrams (or micrograms) per liter of original lake water. It is, of course, possible to prepare the standard curve to yield the amount of the element in sample volume because all standard solutions are expressed as concentration per milliliter. One then converts to concentration in the original water using the following equation:

$$\text{mg (or } \mu\text{g)/L of original water} = \frac{\text{mg (or } \mu\text{g) in sample} \times 1{,}000}{\text{Sample volume (milliliters)}}$$

Errors in the use of the Spectronic 20 that produce invalid results often arise from one of the following:

1. Unmatched test tubes or careless orientation of the tube in the holder.

2. Fluctuation in the light source due to changes in line voltage. This may be partially overcome by the use of an external voltage regulator. However, the analyst may find it best to avoid using the instrument during times of probable variation of demand on the electric lines (4 to 5 PM is often a poor time). The problem of fluctuation is apparent in failure of the instrument to hold a constant setting of zero absorbance.

3. Fatigue of the photocell. This is also evidenced by difficulty in maintaining a constant zero setting.

4. Error in reading the meter. It must be emphasized that the meter is not uniformly accurate over the entire scale. At high absorbancies the scale is crowded in terms of concentration, so that a considerable change in the relative concentration of the substance will cause only a slight change in the position of the needle. At very low absorbancies, slight differences, such as those produced by fingerprints, dust, bubbles, or faulty positioning, will cause a great change in the meter reading. It is preferable to use the center of the scale, between 1 to 0.1 absorbance, by diluting or concentrating the sample, if possible, or by changing the length of the light path.

5. Color or turbidity present in the sample. In many natural waters, turbidity is a major problem. This includes both natural clay turbidity and chemically produced turbidity resulting from the analytical processes that may cause some other substances in the sample to precipitate. Natural turbidity is best controlled by filtration where possible. Photometric compensation may also be used to correct for interference resulting from color, turbidity, or both in *some* samples. This procedure involves the use of a duplicate water sample as a blank in place of the distilled water–reagent blank. This sample blank is carried through all steps of the procedure except the one that causes the development of the measured color. Consequently, natural turbidity or color or arti-

ficially produced turbidity caused by chemical changes up to this point is compensated for by using this sample blank to zero the instrument. Note that this compensation procedure is not universally applicable and that its use must be demonstrated for each analysis. The interferences brought into the analysis by "natural" water causes, such as turbidity or humic acid colors, generally are a greater problem in the blue end of the spectrum because of their higher absorbancies there. It is preferable to work with a wavelength above 500 nm whenever possible.

In an instrument that does not use the entire visible light spectrum, it is necessary to select the best wavelength of light for the particular analysis. In general, this is the wavelength that has the greatest spread of readings between a standard and the blank, that is, the wavelength at which greatest absorbance occurs. However, possible interference from foreign substances in the sample must also be considered, and it may be necessary to select an alternate wavelength free from interferences even though the absorbance may not be as great. For example, in acetone extract of leaf material, chlorophyll *a* absorbs strongly at both 665 nm and 435 nm; however, carotenoid pigments absorb strongly at 480 nm and interfere at 435 nm for chlorophyll. Thus chlorophyll *a* is usually determined at 665 nm. For many analytical procedures, the best wavelength may be found in the literature.

For wavelengths above 625 nm, it is necessary to use a red-sensitive phototube and to insert a red filter. Both of these must be absolutely free from dirt and fingerprints when installed in the instrument.

Calibration of matched tubes

Matched sets of tubes may be purchased, or sets may be made in the laboratory using borosilicate glass test tubes in the following procedure. Purchased sets or laboratory matched sets should be periodically recalibrated.

Apparatus

Spectronic 20 with blue phototube installed and set at 510 nm
25 1-inch test tubes
25 1 cm test tubes
Wood or plastic test tube rack
Cloth towels
Glassware for reagents

Reagents

1% HCl. Add 1 ml concentrated HCl to about 40 ml distilled water in a 100 ml volumetric flask, mix, and make up to mark.

Cobalt chloride solution. Dissolve 2.2 g cobalt chloride in 1% HCl in 100 ml volumetric flask; make up to mark with 1% HCl.

Procedure

1. Wash and dry all test tubes carefully. After washing, handle only by tops, and place in rack or wrap in cloth towel so that glass does not touch glass.

2. Starting with the 1 cm tubes, add about 5 ml of 1% HCl to one tube to serve as a blank.

3. Add similar volumes of cobalt chloride solution to each of the other tubes. Examine each tube to make certain that no air bubbles cling to the inner surface.

4. Wipe each tube clean with a soft cloth or Kimwipe just before inserting tube into the instrument sample holder.

5. With cover of holder closed, set instrument to a wavelength of 510 nm; use left front knob to set needle to the left zero transmittance line (dark-current adjustment). Make certain that right front knob is turned counterclockwise as far as it will go; then wipe tube containing acid blank and insert tube in the holder, being careful to align mark on the tube exactly with marker on the tube holder. Make this alignment while inserting tube; do not turn tube once it is completely down in the holder. Now use right front knob to adjust the gain until the meter reads zero absorbance (100% transmittance) at the right side of the meter scale.

6. Remove the blank and insert the first tube of cobalt

chloride solution, observing all the precautions as for the blank. Record the absorbance for the sample.

7. Remove sample and reinsert blank to check for meter drift; rezero if necessary.

8. Repeat steps 6 and 7 for the remaining tubes, being sure to keep them in order so that they can be identified later.

9. Select the 10 tubes having the narrowest range of absorbancies. Clean these and wrap in cloth towel. These will be your matched set of tubes. You must take all precautions to avoid marring the surface of them in any way, which may destroy the matching qualities. In other words, do not clean them with scouring powder, do not bang together in dishpan or sink, do not lay them on a bare counter surface, and do not place in a wire or metal rack.

10. Repeat this procedure, using the 1-inch tubes, to produce a matched set. If the absorbancies are excessively high with this longer light path, dilute the sample solution with 1% HCl to bring the absorbance back to midscale.

Practice procedures

The following two procedures are included to permit the novice to practice using a simple colored solution ($KMnO_4$) in place of the reagents for specific procedures.

Apparatus

As before with matched tubes

Reagent

Potassium permanganate solution. Dissolve 0.04 g $KMnO_4$ in 1 liter distilled water.

DETERMINATION OF SPECTRAL TRANSMITTANCE CURVE
Procedure

1. Take two clean, dry, 1 cm matched tubes. To the first, add about 5 ml distilled water. To the second, add about 5 ml $KMnO_4$ solution.

2. Set the wavelength scale at 400 nm, and with cover closed, zero left side of scale.

3. Insert blank into holder and adjust with right knob to zero absorbance.

4. Remove blank, insert sample, and record absorbance.

5. Turn down right gain knob, and set wavelength scale to 425 nm. Remove sample, and check left zero with cover closed.

6. Repeat steps 3, 4, and 5 at increments of 25 nm up to 750 nm (change photocell and filter at 625 nm). Plot absorbance against wavelength on linear graph paper. Plot absorbance on the ordinate and wavelength on the abscissa.

7. On the basis of this plot, determine region of rapid change in absorbance with wavelength. Go back and remeasure absorbancies in this region, using 10 nm instead of 25 nm increments. Smooth out your curve on the basis of this better information. Select the best wavelength to determine concentrations of $KMnO_4$ solutions.

DETERMINATION OF A STANDARD CONCENTRATION CURVE
Procedure

1. Use the $KMnO_4$ solution as a 40 mg/L stock solution, and carefully prepare the following dilute concentrations, using best quantitative technique: 30 mg/L, 22.5 mg/L, 15 mg/L, 7.5 mg/L.

2. Set the spectrophotometer at the wavelength determined to be best in the previous procedure; using a distilled water blank, determine absorbance for each concentration, including the stock solution.

3. Plot milligrams of $KMnO_4$ per liter against absorbance on linear graph paper.

pH

The pH of a solution is a measure of its hydrogen ion activity and is the logarithm of the reciprocal of the hydrogen ion concentration. It is important to remember that a change of one pH unit represents a tenfold change in hydrogen ion concentration; for example, a pH of 6.0 has ten times the hydrogen ions of pH 7.0, and pH 5.0 has a hundred times the hydrogen ions of pH 7.0. NOTE: Because of this, it is inappropriate to calculate a mean pH unless you determine the actual hydrogen ion concentrations, compute this mean, and then express it as pH. You may find it more convenient to report a median pH value.

The pH of most natural waters falls in the range of 4.0 to 9.0, and much more often in the range of 6.0 to 8.0. In water, deviation from the neutral pH 7.0 is primarily the result of hydrolysis of salts of strong bases and weak acids or of weak bases and strong acids. However, dissolved gases such as CO_2, H_2S, and NH_3 also have a significant effect. The majority of natural waters have a somewhat alkaline pH because of the presence of carbonate and bicarbonate. The limnologist is interested in pH and its changes, since they may reflect biological activity and changes in natural chemistry of waters, as well as pollution. It is also an important tool for numerous laboratory procedures.

Apparatus

   pH meter
   Celsius thermometer
   Kimwipes
   Beakers

Reagents

   pH buffer solutions in the range of water sample being analyzed
   Distilled water

Procedure for sample collection

The pH of the water may be subject to change during the interval between sampling and determination by reactions such as oxidation or hydrolysis that take place in the bottle. A loss of gases, biological activity, and the absorption of laboratory fumes are all potential sources of change in pH. The sample should be held on ice and analyzed as soon as possible. The analysis is often done in conjunction with the alkalinity determination.

Use of pH meter

Although meters are simple to use and instructions are available for each make, a few precautions will improve the reliability of the determinations.

1. Because water samples are usually relatively unbuffered solutions as compared with many solutions of the chemical laboratory, the meter is often subject to drift and is slow in coming to a steady reading. Readings should not be taken before equilibrium has been established between the electrode and the water system, as evidenced by lack of drift.

2. Temperature exerts two significant effects on pH measurements. Ionization varies with temperature and is an inherent problem in all pH measurements; thus the temperature is usually reported along with the pH. This ionization effect diminishes with an increase in alkalinity. The second effect lies in the pH electrodes themselves, since the actual potential varies with temperature. An adjustment on the instrument may compensate for this variation; however, it is always best to keep the temperature of the sample as near to the temperature of the buffer as possible, and the difference must not exceed $10°$ C.

3. Since the electrodes themselves are the heart of a pH analytical system, care must be taken to see that they are working properly. Care must be taken to avoid scratching the glass electrode against the beaker or wiping with a dirty, gritty cloth or paper. The fiber junction reference electrode must have a free flow of electrolyte through the fiber. If this becomes plugged through dirt, grease, drying out, or other causes, the flow must be returned to a free state or the electrode discarded. An electrode that does not require regular refilling with KCl electrolyte is suspect.

4. Problems also arise from polarization of the elec-

trode system. To avoid polarization, the electrodes *must never be removed* from any solution when the function switch is in any position other than standby or off.

Procedure

1. Prepare a buffer solution with a pH near the expected pH of the water sample. Buffers may be prepared in the laboratory following the instructions in *Standard Methods* or may be purchased in many forms. Since dilute solutions are less stable, it is best to prepare buffers by dilutions of concentrates following the instructions on the bottle. Place 25 ml buffer solution in a small beaker.

2. Open the vent of the reservoir of the fiber junction electrode (usually a plug or sliding sleeve).

3. Rinse the electrodes with distilled water, and blot with soft paper or Kimwipe.

4. Immerse the electrodes in the buffer solution, measure temperature of buffer solution, and set compensator to that temperature.

5. Set function switch to pH, and adjust meter needle to the pH of the buffer. The instrument is now standardized. *Return switch to standby.* NOTE: To check the function of the electrode-potentiometer system, it is desirable to measure the pH of a second buffer of different pH. A small error, approximately 0.1 pH unit, may be encountered. This may be compensated for if the instrument has a "slope" compensated function. It is because of this small error that a buffer of pH near the expected sample pH is preferred.

6. Rinse electrodes in distilled water and blot dry.

7. Immerse electrodes in sample, measure temperature of sample, and set compensator to this temperature.

8. Set function switch to pH, and gently swirl (or use magnetic stirrer at very slow speed) until pH meter needle stops drifting. Read pH from meter scale. *Return switch to standby.* Raise electrodes, rinse with distilled water, and dry. The instrument is now ready for the next sample or for storage.

Storage of the instrument

Close the reservoir vent of the fiber junction electrode to minimize electrolyte loss, and leave electrodes immersed in distilled water.

FREE CARBON DIOXIDE

Carbon dioxide is an end product of decomposing bacteria metabolism and of the respiratory processes of plants and animals. It is also added to the water by the action of natural or pollution acids on bicarbonates. Accurate analysis for this dissolved gas is difficult. The saturation concentration of carbon dioxide is less than 1.1 mg/L at normal water temperatures and atmospheric pressures. Waters are frequently supersaturated with carbon dioxide. As a result, almost any handling process of the water in collection or analysis will result in a loss of the gas.

Because of this analytical problem, several indirect methods of estimating carbon dioxide concentration have been attempted. These are based on the equilibria of the forms of carbonates in water of known temperature, pH, total mineral content, and total alkalinity. The nomographic method in *Standard Methods* is good but requires quantitative knowledge of total filterable residue. Though not a routine determination of limnological surveys, it is not difficult. A second nomographic method (Moore, 1939) is much simpler, requiring quantitative knowledge of only pH and total alkalinity. Because temperature and total mineral content are not considered, the possibility for error in this method is increased. If, after the Moore nomograph is checked against other methods, it is found to provide adequate results for a given source of water, it provides one of the most rapid methods. A third indirect method is the U.S. Geological Survey (Rainwater and Thatcher, 1960) calculation method. For total solids concentrations of less than 500 mg/L and in the pH range of 6.0 to 9.0, a simple calculation based on the bicarbonate equilibria and the hydrogen ion concentration is provided. A final method is direct titration. Completion of this titration occurs at pH 8.3 and is determined either with the pH meter or by the development of a pink color, using a phenolphthalein indicator.

Free carbon dioxide by titration
Apparatus

Titration glassware
Daylight or white light source (if using indicators)
pH meter and accessories (if available)

Reagents

Phenolphthalein indicator solution. See the following section on alkalinity.

Standard sodium carbonate titrant, 0.0454 N. Dissolve 0.602 g anhydrous, primary standard grade Na_2CO_3 that has been oven dried at 140° C overnight in freshly boiled and cooled carbon dioxide–free water, and dilute to the mark in a 250 ml volumetric flask. Store in a rubber-stoppered Pyrex bottle. This solution should not be kept for more than 1 to 2 weeks and, for precise work, should be made fresh daily.

Procedure

Most accurate results are obtained by performing this analysis immediately. If returned to the laboratory, the sealed samples must be placed on ice.

Collect the sample by means of water sampler, and discharge the contents at the bottom of a 100 ml Nessler tube, allowing the sample to overflow several volumes; then withdraw the tube while the sample is flowing. Flick the tube to throw off the water down to the 100 ml mark. Add 10 drops of phenolphthalein indicator. If the sample turns pink, free carbon dioxide is absent. If the sample remains colorless, titrate rapidly with standard sodium carbonate titrant, stirring gently with a glass rod until a faint pink color persists for 30 seconds when viewed through the length of the tube over a white surface. It is wise to have a second cylinder of water to which phenolphthalein has been added, but without titrant, alongside for comparison in determining the faint pink color. For field titration, a 2 to 5 ml serological pipette with small opening may be used as a burette. To check the possibility of carbon dioxide loss during the actual process of titrating, a second sample may be obtained; after adding phenolphthalein, rapidly pour in the full amount of titrant used in the first titration. If the sample turns pink, no carbon dioxide was lost; if it remains colorless, titrate with additional titrant until the pink appears. The result of the second titration is to be accepted.

Calculations

$$\text{mg } CO_2 \text{ per liter} = \frac{A \times N \times 22{,}000}{\text{ml sample}} = \frac{A \times 998.8}{100} = A \times 9.998$$

where

A = ml of titration
N = Normality of sodium carbonate

Table 2. Value of $1.589 \times 10^6 [H^+]$ for each 0.1 pH unit between pH 6.0 and 9.0*

pH	1.589×10^6 [H$^+$]	pH	1.589×10^6 [H$^+$]	pH	1.589×10^6 [H$^+$]
6.0	1.589	7.0	0.159	8.0	0.016
6.1	1.262	7.1	0.126	8.1	0.013
6.2	1.003	7.2	0.100	8.2	0.010
6.3	0.796	7.3	0.080	8.3	0.008
6.4	0.633	7.4	0.063	8.4	0.006
6.5	0.503	7.5	0.050	8.5	0.005
6.6	0.399	7.6	0.040	8.6	0.004
6.7	0.317	7.7	0.032	8.7	0.003
6.8	0.252	7.8	0.025	8.8	0.003
6.9	0.200	7.9	0.020	8.9	0.002
				9.0	0.002

*From Rainwater, F. H., and L. L. Thatcher. 1960. Methods for collection and analysis of water samples. U.S. Geological Water Supply Paper 1454. U.S. Government Printing Office, Washington, D.C.

Free carbon dioxide by calculation

The collection and analysis of samples are the same as described in the section on alkalinity. Alkalinity titration and pH determination must be made as soon as and as accurately as possible. Bicarbonate alkalinity is calculated by dividing total alkalinity as $CaCO_3$ by 0.82.

Calculations

Milligrams of carbon dioxide per liter can be calculated in the pH range of 6.0 to 9.0 using the following:

$$\text{mg } CO_2 \text{ per liter} = 1.589 \times 10^6 \left[H^+\right] \times \text{mg/L alkalinity as } HCO_3$$

or

$$\text{mg } CO_2 \text{ per liter} = 1.589 \times 10^6 \left[H^+\right] \times \frac{\text{mg/L total alkalinity as } CaCO_3}{0.82}$$

The value of the term $1.589 \times 10^6 \left[H^+\right]$ is given in Table 2 for each 0.1 pH unit between 6.0 and 9.0.

Free carbon dioxide by Moore nomograph

Using accurate pH and total alkalinity, estimate milligrams of free carbon dioxide per liter from the nomograph (Fig. 12).

ALKALINITY

The alkalinity of water is its capacity to accept protons; stated another way, it is the quantity and kinds of compounds present that collectively shift the pH to the alkaline side of neutrality. Although the alkalinity of natural waters is generally the result of bicarbonates, it is usually expressed in terms of calcium carbonate. Three kinds of alkalinity are indicated: hydroxide (OH^-), normal carbonate ($CO_3^=$), and bicarbonate (HCO_3^-), the three are summed as total alkalinity. Carbonates and bicarbonates are common to most waters because carbonate minerals are abundant in nature and because contribution to alkalinity by hydroxides is rare in nature. The presence of hydroxides can usually be attributed to water

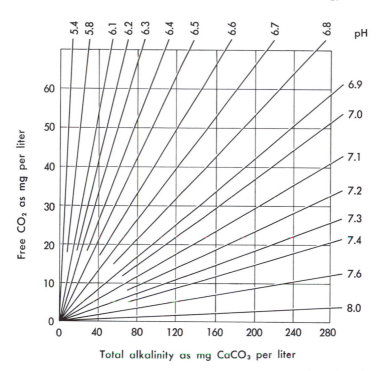

Fig. 12. Moore's nomograph for determination of free carbon dioxide concentration from pH and total alkalinity data.

treatment or to contamination. Expected total alkalinities in nature usually range from 20 to 200 mg/L.

Principle of determination

Five conditions of alkalinity are possible in the sample: carbonate, bicarbonate, or hydroxide alone, or a combination of carbonate and hydroxide or of carbonate and bicarbonate. Hydroxide and bicarbonate are not found together.

Alkalinity is determined by titrating the water sample with a standard solution of strong acid. The equivalency or end points of the titration are selected as the inflection points in the titration of sodium carbonate with H_2SO_4. These points will vary slightly with temperature, ionic concentration, and free carbon dioxide concentration. The end points may be determined empirically by titration and are the pH where the derivative of $\Delta pH/\Delta ml$ titrant is greatest. However, the points are usually taken as pH 8.3 for the carbonate end point and pH 4.5 for the bicarbonate end point. The following reactions occur in titrations (Rainwater and Thatcher, 1960):

$CO_3^= + H^+ \rightarrow HCO_3^-$ (titration to pH 8.3)

HCO_3^- (from $CO_3^=$) $+ H^+ \rightarrow H_2O + CO_2$

HCO_3^- (natural) $+ H^+ \rightarrow H_2O + CO_2$ (titration to pH 4.5)

Apparatus

Titration glassware
Magnetic stirrer if available
Daylight light source

Reagents

Carbon dioxide–free water. Boil for 15 minutes and cool to room temperature; pH must be greater than 6.0.

Phenolphthalein indicator solution. Dissolve 0.5 g phenolphthalein in 50 ml 95% ethyl or isopropyl alcohol. Add 50 ml distilled water, mix, and add weak sodium hydroxide solution drop by drop until a very faint pink color appears.

Stock sulfuric acid solution. Dilute 3 ml concentrated H_2SO_4 to 1 liter; this gives approximately 0.1 N solution.

Standard sulfuric acid solution. Dilute 200 ml 0.1 N stock solution to 1,000 ml with carbon dioxide–free water to give 0.02 N solution. For accurate work this must be standardized against Na_2CO_3 (American Public Health Association, 1976).

Methyl orange indicator solution. Dissolve 0.5 g methyl orange in 1 liter distilled water.

Bromcresol green–methyl red indicator solution. Dissolve 0.02 g methyl red and 0.10 bromcresol green in 100 ml 95% alcohol.

Procedure

Collect 250 ml or greater samples without aeration in glass or polyethylene bottles. Alkalinity is susceptible to change between time of collection and analysis. The main problem is loss of carbon dioxide from solution, which results in the conversion of bicarbonate to carbonate. Total alkalinity values are probably somewhat more stable than the relative values of the common alkalinity components. Analysis must be done within a few hours of collection, and the sample must be kept sealed until analysis is ready to begin; it must not be filtered, diluted, or altered in any way. Gently pour out 100 ml of sample and place in a 250 ml Erlenmeyer flask (if indicators are to be used) or a 200 to 250 ml beaker (if pH meter is to be used).

Titration with pH meter. Place the prepared sample on a magnetic stirrer and add spin bar. Place prepared pH electrodes in sample. Turn on stirrer at slow speed (do not agitate surface of liquid), and record sample pH when meter drift stops.

Titrate with standard acid solution, recording titrant volume at pH 8.3 and pH 4.5. If a magnetic stirrer is not available, swirl the sample gently by hand.

Titration with indicators. Although a meter is preferable, indicators for determination of pH end points may be used. Add 4 drops of phenolphthalein indicator to the sample. If a pink color appears, hydroxide or normal carbonate is present. Titrate with standard sulfuric acid until color disappears, and record milliliters of acid used. If no pink color appears, or after phenolphthalein titration, add 2 drops of methyl orange indicator, and titrate to a faint orange. This is a difficult end point and usually requires some experience to recognize. Overtitration (to pH 4 or below) produces a pink color. As an alternate method, one can substitute bromcresol green–methyl red indicator for the methyl orange and titrate until the blue changes to a *clear* pink (a gray-pink appears before the final change). It is wise to practice recognizing this end point by using meter and indicator together.

Calculations

$$\text{Phenolphthalein alkalinity as mg } CaCO_3 \text{ per liter} = \frac{A \times N \times 50{,}000}{\text{ml sample}}$$

$$\text{Total alkalinity as mg } CaCO_3 \text{ per liter} = \frac{B \times N \times 50{,}000}{\text{ml sample}}$$

where

A = ml titration to pH 8.3 (phenolphthalein end point)
B = ml total titration from start to pH 4.5 (methyl orange end point)
N = Normality of acid

NOTE: When titrating a 100 ml sample with 0.02 N H_2SO_4, the equations are

$$\text{Phenolphthalein alkalinity as mg } CaCO_3 \text{ per liter} = A \times 10$$

and

$$\text{Total alkalinity as mg } CaCO_3 \text{ per liter} = B \times 10$$

Calculation of alkalinity relationships

Results obtained from the titrations offer a means of classifying the three principal forms of alkalinity present. Mathematical conversion of titration results may be made using Table 3.

Bicarbonate alkalinity as mg HCO_3 per liter is determined by dividing total alkalinity by 0.82.

Table 3. Alkalinity relationships*

Result of titration	OH alkalinity as CaCO$_3$	CO$_3$ alkalinity as CaCO$_3$	HCO$_3$ alkalinity as CaCO$_3$
$P = 0$	0	0	T
$P < \frac{1}{2} T$	0	2P	T−2P
$P = \frac{1}{2} T$	0	2P	0
$P > \frac{1}{2} T$	2P−T	2(T−P)	0
$P = T$	T	0	0

*P, Phenolphthalein titration; T, total titration.

HARDNESS

Water that contains salts of calcium and magnesium and, to a lesser extent, other polyvalent metals, such as iron, aluminum, and manganese, and that requires large amounts of soap to lather or that, on evaporation, forms a deposit on the container is referred to as hard water. In early methodology, hardness was measured as the capacity of water to precipitate soap when a liquid soap solution was shaken with the water sample to form a lather persisting for 5 minutes. Hardness is now best defined as a characteristic of water representing the total concentration of calcium and magnesium ions expressed as milligrams of $CaCO_3$ per liter. When other ions are present in insignificant amounts, the hardness will be equal to or less than the sum of the carbonate and bicarbonate alkalinities and is termed *carbonate hardness*. If the hardness exceeds the sum of these alkalinities, the presence of other ions is indicated, and the excess is expressed as *noncarbonate hardness*. Hardness determination is thus corroboratory to the findings of alkalinity; however, when used in conjunction with calcium determinations, it allows for the calculation of magnesium concentration, circumventing the laboratory analysis for this ion. Hard water is not generally considered harmful to man (except as a possible cause of kidney stones); within limitations, the ions present are necessary for normal plant and animal survival and growth. Evidence suggests that hardness may affect tolerance of fishes to toxic metals.

What actually constitutes "hard water" has been variously described. One reasonable classification is that by Brown, Skougstad, and Fishman (1970) for hardness as mg per liter $CaCO_3$ and is as follows: 0 to 60 = soft, 61 to 120 = moderately hard, 121 to 180 = hard, and 181 and above = very hard.

The Eriochrome Black T indicator used in this titration produces a red color in the presence of calcium and magnesium ions. The disodium salt of EDTA forms a stable, colorless complex with these ions, effectively removing them from solution. Thus by titration of the sample containing the indicator with EDTA, the calcium and magnesium ions are quantitatively removed. When all ions are removed, the indicator changes to a bright blue color.

Apparatus

Titration glassware
Magnetic stirrer if available
Daylight light source

Reagents

Buffer solution. Use commercially prepared buffer as directed, or prepare per instructions in *Standard Methods.*

Indicator. Mix 0.8 g Eriochrome Black T dye and 100 g NaCl to prepare a dry powder mixture.

Standard EDTA titrant, 0.01 M. Dissolve 0.3723 g Na_2EDTA-dihydrate in distilled water and dilute to 100 ml. Check by titrating against a standard calcium solution: 1.00 ml = 1.00 mg $CaCO_3$ = 0.4008 mg Ca.

Standard calcium solution. Weigh 1.000 g anhydrous calcium carbonate powder, primary standard grade, into a 500 ml Erlenmeyer flask. Add slowly one volume HCl diluted with an equal volume of distilled water (1 + 1 HCl) until all the $CaCO_3$ has dissolved. Add 200 ml distilled water, and boil for a few minutes to expel CO_2. Cool, and adjust to pH 5.0 with either NH_4OH (about 3 N) or 1 + 1 HCl. Transfer to a 1-liter volumetric flask, washing out the Erlenmeyer flask several times with distilled water and adding to volumetric flask. Then dilute to mark with distilled water. (1.00 ml = 1.00 mg $CaCO_3$ = 0.4008 mg Ca.)

Procedure

1. Dilute 25 ml of sample to about 50 ml with distilled water in titration flask.

2. Add 1 to 2 ml of buffer solution to bring pH to 10.0 or 10.1.

3. Add approximately 0.1 g indicator powder (just a bit on the tip of a spoon or spatula).

4. Titrate with EDTA over a white surface with daylight or white light. Stir continuously until the last red

tinge disappears. Add the last drops slowly, allowing about 5 seconds between drops. The entire duration of titration should not exceed 5 minutes and should not require more than 15 ml of titrant. If more titrant than this is used, take a smaller aliquot and repeat titration. An indistinct end point suggests interference and calls for an inhibitor after step 2. Faulty (old) indicator powder also produces an indistinct end point.

Calculations

$$\text{EDTA hardness as mg } CaCO_3 \text{ per liter} = \frac{A \times B \times 1,000}{\text{ml of sample}}$$

where

A = ml titration
B = mg $CaCO_3$ equivalent to 1.00 ml EDTA titrant

NOTE: When titrating a 25.0 ml sample with 0.01 M EDTA titrant, use the following equation:

$$\text{EDTA hardness as mg } CaCO_3 \text{ per liter} = A \times 40$$

Express results as either total hardness, carbonate hardness, or noncarbonate hardness, as appropriate.

CALCIUM

Calcium can be leached from practically all rocks but is much more prevalent in waters from regions with deposits of limestone, dolomite, and gypsum. Regions in which granite or siliceous sand predominates have very low calcium concentrations in the waters. Concentrations in waters from limestone areas range from about 30 to 100 mg/L. Calcium is important to the biological productivity of waters. It is a major constituent of the cell walls of higher aquatic plants and in supporting structures of many animals such as the bony tissues of fishes and the shells of mollusks. Calcium is considered as a micronutrient for most algae, in which it plays a more functional role in membrane selective ion transport. Waters with a concentration of 10 mg or less per liter are usually oligotrophic, whereas waters with 25 mg or more per liter are usually distinctly eutrophic.

The EDTA titrant used in this calcium determination is the same as for water hardness and reacts with both calcium and magnesium. To determine calcium only, the pH is made sufficiently high so that the magnesium precipitates as magnesium hydroxide, and an indicator specific for calcium is used.

Apparatus

Titration glassware
Magnetic stirrer if available
Daylight or white light source

Reagents

Sodium hydroxide, 1 N. Dissolve 4 g NaOH in distilled water, and when cool, dilute to 100 ml.

Murexide indicator. Grind together in a mortar 0.2 g powdered dye and 100 g NaCl. Store in tightly stoppered bottle.

Standard EDTA titrant, 0.01 M. Same as in hardness determination. (1.00 ml = 0.4008 mg Ca.)

Procedure

1. Take a sample that contains less than 10 mg Ca. Usually a 50 ml water sample is correct, but if total alkalinity is greater than 250 mg/L, it probably will be better to take a smaller aliquot and dilute to 50 ml with distilled water.

2. Add 1 to 2 ml NaOH solution to produce a pH of 13 to 14. Stir.

3. Add about 0.2 g indicator powder. The color change is from pink to purple on titration.

4. With continuous stirring, titrate slowly over a white surface with the EDTA titrant. Since this is a gradual color change, the end point recognition is facilitated by preparing a reference end point by adding NaOH, indicator, and 1 or 2 ml EDTA to 50 ml distilled water.

Calculations

$$mg\ Ca\ per\ liter = \frac{A \times B \times 400.8}{ml\ of\ sample}$$

where

A = ml titration for sample
B = mg $CaCO_3$ equivalent to 1.00 ml EDTA titrant

NOTE: The equation for titrating a 50 ml sample with 0.01 M EDTA titrant follows:

$$mg\ Ca\ per\ liter = A \times 8.016$$

MAGNESIUM

Magnesium in natural waters comes mainly from the leaching of igneous and carbonate rocks. In areas where these sources are common, magnesium concentrations in water often range from 5 to 50 mg/L. Magnesium is related to water hardness in the same manner as calcium. This element is of concern to the limnologist in its role as an essential nutrient in plant growth and development, especially as related to its function in the chlorophyll molecule.

Procedure

There are several analytical methods for magnesium, but if both calcium concentration and hardness are known, magnesium concentration may be calculated from them (Rainwater and Thatcher, 1960).

Milliequivalents of hardness per liter are calculated from milligrams of hardness per liter. The milliequivalents of calcium per liter are subtracted from this, and the difference is multiplied by the equivalent weight of magnesium. This gives magnesium in milligrams per liter.

Calculations

mEq hardness per liter = mg hardness per liter × 0.01998
mEq Ca^{+2} per liter = mg Ca^{+2} per liter × 0.0499
mg Mg^{+2} per liter = 12.16 × (mEq hardness per liter − mEq Ca^{+2} per liter)

DISSOLVED OXYGEN

The dissolved oxygen (DO) content of waters results from (1) the photosynthetic and respiratory activities of the biota in the open water, the benthos, and the aufwuchs; and (2) the diffusion gradient at the air-water interface and distribution by wind-driven mixing. Generally 3 mg DO or less per liter is stressful to aquatic vertebrates and most other aquatic life.

The azide modification of the Winkler method is applicable to most natural waters; however, nitrites, ferrous and ferric iron, sulfites, sulfides, polythionates, and other sulfur compounds interfere with the correct determination of DO when the Winkler reagents are used. Both nitrites and ferric iron liberate additional iodine from the potassium iodide during the course of the procedure so that the final titration is too high. Ferrous iron, sulfites, and organic substances remove some of the iodine liberated from the potassium iodide by the manganic compounds so that the final titration is too low. If these interferences are suspected, orientation tests to demonstrate their presence or absence may be conducted (Ellis, Westfall, and Ellis, 1948). Any readily oxidizable or reducible substances may cause interference. Oxidizable substances cause low DO results, whereas reducible substances cause high results. In this method, the sodium azide eliminates the interference of nitrite, and the addition of a simple initial potassium fluoride step overcomes the effect of ferric iron up to 200 mg iron per liter. In the absence of interference, the precision of this method is ±0.05 mg per liter, except at low concentrations.

Principle of determination

This method depends on the formation of a loose precipitate (floc) of manganous hydroxide. The oxygen dissolved in the water is rapidly absorbed by the manganous hydroxide, forming a higher oxide, which may be in the following form:

$$MnSO_4 + 2KOH \rightarrow Mn(OH)_2 + K_2SO_4$$
$$2Mn(OH)_2 + O_2 \rightarrow 2MnO(OH)_2$$

As it settles, the $Mn(OH)_2$ floc acts as a "gathering" agent for the DO. On acidification in the presence of iodide, iodine is released in a quantity equivalent to the DO present:

$$MnO(OH)_2 + 2KI + H_2O \rightarrow Mn(OH)_2 + I_2 + 2KOH$$

The liberated iodine is then titrated with a standard sodium thiosulfate or PAO solution, with starch used as the indicator (Rainwater and Thatcher, 1960):

$$I_2 + 2S_2O_3^{-2} \rightarrow S_4O_6^{-2} + 2I^{-1}$$

Reagents

Manganese sulfate solution. Dissolve 120 g $MnSO_4 \cdot 4H_2O$, or 100 g $MnSO_4 \cdot 2H_2O$, or 91 g $MnSO_4 \cdot H_2O$ in distilled water; filter, and dilute to 250 ml (do not store in glass-stoppered bottle).

Alkali-iodine-azide reagent. Dissolve 125 g sodium hydroxide and 33.75 g sodium iodide in distilled water, and dilute to 250 ml. To this solution add 2.5 g sodium azide *(poison)* that has been dissolved in 10 ml distilled water.

Starch solution. Prepare an emulsion of 6 g starch in a beaker with a *small* amount of distilled water. Pour this emulsion in 1 liter of distilled water, boil for a few minutes, and let settle overnight. Save the clear supernate, and preserve by adding 1.25 g salicylic acid (it is best to store the stock bottle under refrigeration).

Sulfuric acid. Concentrated.

Sodium thiosulfate stock solution, 0.10 N. Dissolve 12.41 g $Na_2S_2O_3 \cdot 5H_2O$ in boiled and cooled distilled water; dilute to 500 ml. Preserve by adding 2 g borax ($Na_2B_4O_7 \cdot 10H_2O$), or 0.5 g NaOH or 2.5 ml chloroform.

Standard sodium thiosulfate titrant, 0.0125 N. Dilute 125.0 ml sodium thiosulfate stock solution, 0.10 N, to 1,000 ml with distilled water. Preserve by adding either 3 g borax, 0.4 g NaOH, or 5 ml chloroform.

This titrant solution must be standardized by comparison with standard potassium iodate (KIO_3). Place 25.00 ml of the standard KIO_3 in a 250 ml Erlenmeyer flask, and add 75 ml demineralized water. Dissolve 2 g potassium iodide (KI) in this solution. After all of the potassium iodide is dissolved, add 10 ml of 3.6 M H_2SO_4 (carefully dilute 200 ml concentrated H_2SO_4 to 1 liter). Allow this to stand at room temperature in the dark for 2 to 5 minutes. Titrate with the sodium thiosulfate, adding 2 ml starch indicator when the titration has reached a pale straw color. Continue titration until blue color disappears. The actual normality of the standard sodium thiosulfate may be calculated as follows:

$$N \text{ of } Na_2S_2O_3 = \frac{25}{ml\ Na_2S_2O_3\ used\ in\ titration} \times 0.0125$$

The standard sodium thiosulfate may then be adjusted to exactly a 1:1 equivalence by the addition of either stock thiosulfate solution or distilled water. An alternative is to consider the result of the standardization as a correction factor (K) and correct the empirically determined DO values by multiplying by K. NOTE: A widely used alternate titrant is phenylarsine oxide (PAO) supplied by the Hach Chemical Co. (Appendix 2) as a 0.025 N solution. This solution is more stable than the sodium thiosulfate titrant. Since PAO at 0.025 N is twice as concentrated, one should substitute a 200 ml sample for titration for the 100 ml sample called for in step 7 of the procedure.

Standard potassium iodate solution, 0.0125 N. Although it is probably best for most inexperienced workers to purchase stock potassium iodate solutions of 0.1 N and make the appropriate dilutions, it is possible to prepare a standard iodate solution as follows. Oven dry approximately 1 g KIO_3 for 1 hour at 105° C. Dissolve 0.4459 g in demineralized water, and dilute to 1,000 ml. This standard potassium iodate solution must be made up fresh daily.

Procedure

1. Insert the delivery tube of the water sampler to the bottom of a glass-stoppered bottle. (The 300 ml biochemical oxygen demand [BOD] bottles with tapered glass stoppers are preferred, but any 125 or 250 to 300 ml bottle may be used.) Allow the water to flow into the bottom of the bottle until it has overflowed two or three times its volume, making sure that no air bubbles are retained. With flow continuing, remove the tube from the bottle, and immediately stopper the bottle without trapping any air bubbles. This is easy with BOD bottles, but more difficult with flat stoppers. Using flat stoppers, tip the bottle

slightly, partially insert the stopper, and then with a quick twisting motion, seat the stopper.

2. If percent saturation is to be computed, water temperature must be taken.

3. Add 2 ml $MnSO_4$ solution to the sample in a 300 ml bottle by inserting tip of pipette just below water surface.

4. Add 2 ml alkaline-iodide-azide reagent in same manner as for $MnSO_4$.

5. Stopper as described above, and mix by inverting ten to twenty times. Allow the floc to settle until one third of the bottle is clear, then mix again. Allow the floc to settle a second time.

6. When at least one third of the bottle is clear, carefully remove the stopper and add 2 ml concentrated H_2SO_4 by allowing the acid to flow down the neck of the bottle. Restopper and invert gently until solution is complete. In the absence of interferences, the treated sample may be held for up to 3 days under refrigeration before titration.

7. Titration: mix by inversion and pour out 100 ml of sample for titration. For highest precision, this volume taken for titration should correspond to 100 ml of sample water before the addition of reagents. Thus the volume taken for titration using the quantities of reagents described above is 101.5 ml. (For method of calculation see American Public Health Association, 1976, or Ellis, Westfall, and Ellis, 1948.) For routine work this correction is not necessary.

8. Titrate with 0.0125 N thiosulfate solution to a very pale yellow color, add 1 ml starch solution (if the sample turns black instead of blue, you added the starch too soon; this may result in slight error), and continue titration until the first disappearance of the blue color. Record milliliters of thiosulfate. Disregard any reappearance of blue color. If the color reappears within 1 to 2 minutes, it is an indication of an accurate titration, since if the end point is much overrun, the recoloration does not occur. If the end point is overrun, take an additional

Table 4. Solubility of oxygen in pure water exposed to water-saturated air at mean sea level pressure of 760 mm Hg*

Temperature (° C)	DO (mg/liter)	Temperature (° C)	DO (mg/liter)
0	14.16	18	9.18
1	13.77	19	9.01
2	13.40	20	8.84
3	13.05	21	8.68
4	12.70	22	8.53
5	12.37	23	8.38
6	12.06	24	8.25
7	11.76	25	8.11
8	11.47	26	7.99
9	11.19	27	7.86
10	10.92	28	7.75
11	10.67	29	7.64
12	10.43	30	7.53
13	10.20	31	7.42
14	9.98	32	7.32
15	9.76	33	7.22
16	9.56	34	7.13
17	9.37	35	7.04

*Modified from Hutchinson, G. E. 1957. A treatise on limnology, vol. 1. John Wiley & Sons, Inc., New York.

10 ml of sample and add to the first 100 ml; this should cause a blue color to reappear. Then titrate to end point and mathematically correct for the larger sample.

Calculations

Since 1 ml of 0.0125 N thiosulfate is equivalent to 0.1 mg DO, each milliliter of sodium thiosulfate titrant used is equivalent to 1 mg DO per liter when a volume of 100 ml is titrated.

To express the results as percent saturation, the solubility data from Table 4 may be used.

Lower solubilities are found in natural waters because of dissolved salts, but these differences are not significant in nonpolluted fresh waters. To calculate the solubility of DO at any other atmospheric pressure, the following formula may be used, provided the water temperature is below 25° C.

$$S' = S\left(\frac{P}{760}\right)$$

where

S′ = Solubility in mg/L
S = Solubility at 760 mm Hg (760 mm = 29.92 inches)
P = Atmospheric pressure at time of sample collection

PHOSPHORUS

Phosphorus in waters is present in several soluble and particulate forms, including organically bound phosphorus, inorganic polyphosphates, and inorganic orthophosphates. These orthophosphates are usually ions of phosphoric acid. At pH concentrations of most natural waters (less than pH 9.0), the dihydrogen and monohydrogen phosphate ions are prevalent, although analytical techniques usually employed do not distinguish between ionic states, and all inorganic phosphate is usually considered as PO_4.

Because phosphorus is a biologically active element, it cycles through many states in the aquatic ecosystem, and its concentration in any one state depends on the degree of metabolic synthesis or decomposition occurring in that system; for example, lower concentrations are expected at times of high synthetic activity. Although the natural source of phosphorus to waters is from leaching of phosphate-bearing rocks and from organic matter decomposition, additional sources are found in manmade fertilizers, domestic sewage, and detergents. Phosphorus is lost from the water by chemical precipitation to the sediments and by adsorption on clays.

Phosphorus is the element most likely to cause a stimulation of plant production and is thus often measured in relation to excessive production and problems of eutrophication. According to Liebig's "law of the minimum," the element present in the lowest concentration relative to demand is the element limiting the process at that time. The average ratio of three important elements in aquatic macrophytes and algae is 40C:7N:1P per 100 dry weight. It is often desirable to measure the available concentrations and ratios of these elements in the water. Although phosphorus is most often found to be the element in "short supply," occasionally nitrogen and, in rare and unusual conditions, carbon limit plant production. Confirmation that a particular element, such as phosphorus, is limiting may be obtained from bioassays in which the plant community (or culture species in the water) is "spiked" by an addition of the suspected limiting nutrients, either singly or in combination, and any resulting increase in production measured. Total phosphorus concentrations of nonpolluted waters are usually less than 0.1 mg P per liter (100 μg P per liter), and inorganic (orthophosphate) soluble phosphorus is often less than 0.01 mg P per liter.

The following method of Murphy and Riley (1962) specifically measures the orthophosphate (PO_4-P) concentration. If either polyphosphates or organic phosphorus determinations are desired, these may be made by differential hydrolysis to the orthophosphate state, and then the following method employed. Because of the possible hydrolysis of other forms during the Murphy-Riley procedure with consequent overestimations of the inorganic PO_4-P concentration, some prefer to identify the results as soluble reactive phosphorus (SRP). Minimum detectable concentration by the Murphy-Riley method is approximately 20 μg/L (ppb) when a 1-inch sample tube is employed with a Spectronic 20. For unpolluted waters in many portions of the United States, this is inadequate sensitivity. In this case, an additional extraction step (modified from Stephens, 1963) may be employed to bring the sensitivity to the 2 ppb range.

It must be stressed that when one is analyzing for phosphorus in the ppb range, laboratory contamination from dust and detergents will often produce higher concentrations than the lake or river water itself. Phosphorus-containing detergents should be banned from the laboratory in which phosphorus analyses are made. All glassware to be used for phosphorus determination must be acid washed and preferably stored submerged in a dilute acid (1% to 5%) solution. This glassware should never be used for any other types of laboratory analysis. Reagent blanks on glass-redistilled water should be carried through with each analysis to confirm the lack of contamination.

Because of the reported possibility of phosphorus adsorption onto polyethylene, samples for phosphorus analysis should be collected in acid-washed Pyrex bottles. Samples should be refrigerated immediately, and analysis should be completed within a few hours. If analysis must be delayed, any filtration must be done immediately, and the samples must be stored by freezing.

Soluble reactive phosphorus (orthophosphate phosphorus)

Apparatus

Spectronic 20 with red phototube and filter set at 880 nm (690 nm may also be used but tends to be less sensitive)

1-inch matched test tubes

Acid-washed glassware (Erlenmeyer flasks, graduates, pipettes)

Reagents

Glass-distilled water. Use only glass-redistilled water stored in polyethylene bottles for all reagents and blanks.

Phenolphthalein indicator. See section on alkalinity.

Sulfuric acid solution. Slowly add 7 ml concentrated H_2SO_4 to distilled water, and dilute to 50 ml.

Ammonium molybdate reagent. Dissolve 4 g $(NH_4)_6Mo_7O_{24}$ · $4H_2O$ in 100 ml distilled water. Refrigerate in polyethylene bottle for storage.

Potassium antimonyl tartrate. Dissolve 1.3715 g potassium antimonyl tartrate (K[SbO]$C_4H_4O_6$ · $0.5H_2O$) in 400 ml distilled water, and make to volume in a 500 ml volumetric flask. Refrigerate in a polyethylene bottle for storage.

Ascorbic acid solution. Dissolve 1.76 g ascorbic acid in 100 ml distilled water. Freeze in a polyethylene bottle for storage. This may be thawed and refrozen repeatedly.

Strong acid solution. Slowly add 30 ml concentrated H_2SO_4 to 60 ml distilled water.

Mixed reagent. Add 5 ml potassium antimonyl tartrate solution to 50 ml sulfuric acid solution and mix. Add 15 ml ammonium molybdate solution and mix. Add 30 ml ascorbic acid solution and mix. This reagent should be prepared fresh for each set of analyses.

Stock phosphorus solution. Although it is preferable for the occasional worker to purchase commercially prepared stock solutions, stock and standard solutions may be prepared as follows. A stock solution in which 1.00 ml equals 0.05 mg phosphorus is prepared by dissolving 0.2197 g potassium dihydrogen phosphate (KH_2PO_4) (oven dried at 105° C for 1 hour) in distilled water. Dilute this to 1.0 liter. Add 1 ml chloroform, and store in the dark under refrigeration. This stock solution will remain stable for months.

Standard solution (1.00 ml = 0.5 μg P). Dilute 10.0 ml of stock phosphorus solution to 1.0 liter with glass-distilled water. This standard solution should not be stored for more than a few days.

Procedure

Water should be settled free of suspended matter, or it may be filtered, if at all turbid, using either membrane, glass fiber, or paper filters that have been shown to be free of phosphorus. Rinse filter with acid rinse, and follow with glass-distilled water rinse before filtering sample. A blank of glass-distilled water should be run through the same procedure so that contamination at any step of the procedure will show up as a blue color in the blank.

Carry two different standards of approximately the expected phosphate concentration through with the samples.

1. Add 1 drop of phenolphthalein indicator to a 50 ml sample. If a pink color appears, add strong acid solution drop by drop until the color disappears; do not add more

than 5 drops of acid. If more than 5 drops are required, take a 40 ml sample, discharge pink color with acid; then make up to 100 ml with distilled water, and make the necessary correction in your calculations for the smaller volume of sample.

2a. Add 8.0 ml of the mixed reagent to the sample, and mix thoroughly. If phosphorus is present, a blue color will develop. After 10 minutes measure the absorbance of each sample at 880 or 690 nm in the spectrophotometer, using the reagent blank as the reference solution. Determine concentration of phosphorus from a standard phosphorus calibration curve. If the two standards run with the samples do not agree with this standard curve, the reagents have changed and a new curve must be made.

2b. For very low concentrations, the blue color may be concentrated by extracting it out of a large volume of water into a small volume of organic solvent as follows:

Place 200 ml of water sample (or standard) in a 250 ml separatory funnel. Add 40 ml of mixed reagent and mix well. Wait 20 minutes, and add 10.0 ml reagent grade butyl acetate. Mix by inversion for 3 minutes. Allow layers to separate, then draw off and discard the lower aqueous layer. Measure the absorbance of the blue butyl acetate extract at 880 or 690 nm. Always use a reagent blank as the reference solution. Determine phosphorus

Table 5. Volume of standard solution (1.00 ml = 0.5 μg P) to be made up to 50 ml with distilled water, with the resultant concentration

Milliliters of standard solution	Phosphorus concentration (μg/L)
0.5	5
1.0	10
3.0	30
5.0	50
10.0	100

concentration from a standard calibration curve prepared using the same procedure.

Standard curve for phosphorus concentration of the original water sample. For many natural waters, the dilutions shown in Table 5 have been found suitable.

Total phosphorus

All forms of phosphorus are oxidized in the raw water or in filtered water (total soluble phosphorus) to inorganic phosphate using the persulfate digestion method of Menzel and Corwin (1965). The resulting phosphate is analyzed by the previously described Murphy and Riley method.

Apparatus

As for soluble reactive phosphorus
Autoclave (or pressure cooker) or boiling water bath
Acid-washed glassware (Erlenmeyer flasks, graduated cylinders)
Aluminum foil

Reagents

As for soluble reactive phosphorus
Digestion reagent. Prepare a 5% solution (w/v) of potassium persulfate ($K_2S_2O_8$) by dissolving 50 g of the reagent in glass-distilled water and diluting to 1 liter.

Procedure

Separate standards must be prepared and carried through all steps as the samples.

1. Place 50 ml of sample in an Erlenmeyer flask, and add 8.0 ml digestion reagent. Cap tightly with aluminum foil. Swirl to mix.

2. Digest the organic matter in the sample by autoclaving at 15 psi (1 Kg/cm^2) for 30 minutes (a pressure cooker may be substituted for the autoclave). The digestion may also be carried out in a boiling water bath for 1 hour.

3. Cool the sample and make up to 60 ml with glass-distilled water.

4. Proceed with the analysis of SRP on the 60 ml as given before, using step 2a but increasing the volume of mixed reagent to 10 ml.

NITROGEN

Nitrogen is present in aquatic ecosystems in several forms. The great reservoir of nitrogen, the air, is available to aquatic organisms only through the activity of a relatively few species of bacteria and blue-green algae that are capable of "fixing" this source into a form available to the remainder of the biota. The nitrogen cycle is thus a double cycle, that is, one cycle of oxidation and reduction of nitrogen by the typical plants, animals, and decomposers and the larger cycle, which is coupled to the smaller by the action of nitrogen-fixing organisms and denitrifying bacteria. Nitrogen is occasionally limiting to plant production and is probably second only to phosphorus in this role for most aquatic ecosystems (pp. 77 and 78). In the water, the forms of common analytical interest include dissolved and particulate organic nitrogen, which are various reduced forms ranging from simple amines to complex proteins in both living and dead tissues and tissue products; ammonia, which is the most reduced form and is the product of organic decomposition; and nitrate and nitrite, which are oxidized forms resulting from the nitrification (bacterial oxidation) of ammonia. Organic nitrogen is usually determined by the Kjeldahl digestion procedure or by photooxidation of the organic compounds by ultraviolet radiation. Either of these procedures for organic nitrogen requires relatively complex equipment and will not be described here. Good descriptions are those of *Standard Methods* (American Public Health Association, 1976) for Kjeldahl digestion and of Stainton, Capel, and Armstrong (1974) for ultraviolet oxidation.

Ammonia nitrogen

Ammonia is usually present in low (less than 1 mg/L) quantities in nonpolluted, well oxygenated water, but may reach 5 to 10 mg/L in the anaerobic hypolimnion of a eutrophic lake. High concentrations of ammonia when present as the undissociated NH_4OH are toxic. The relative proportion of the toxic NH_4OH increases with pH.

Apparatus

Spectronic 20 with red phototube and filter and set at 640 nm
Matched 1-inch test tubes
50 ml stoppered graduated cylinders or other stoppered glass bottles of at least 60 ml total capacity

Reagents

Hypochlorite solution. Fresh commercial household bleach.

Alkaline citrate solution. Dissolve 20 g sodium citrate ($Na_3C_6H_5O_7 \cdot 2H_2O$) and 1 g sodium hydroxide (NaOH) in demineralized water, and dilute to 100.0 ml with demineralized water. NOTE: Be certain demineralizer is functioning properly to remove traces of ammonia.

Nitroprusside catalyst reagent. Dissolve 1 g sodium nitroprusside (sodium nitroferricyanide) ($Na_2Fe[CN]_5NO \cdot 2H_2O$) in demineralized water, and dilute to 200.0 ml with demineralized water. This is the most unstable of the reagents. Deterioration is indicated by excessively high reagent blanks.

Phenol reagent. Dissolve 20 g phenol in 95 % ethyl alcohol, and dilute to 200.0 ml with alcohol.

Table 6. Volume of standard solutton (1.00 ml = 0.010 mg NH_3-N) to be diluted to 100 ml, with the resultant concentration

Milliliters of standard solution	Ammonia nitrogen concentration (mg/L)
0.5	0.05
1.0	0.10
2.5	0.25
5.0	0.50
10.0	1.00

Oxidizing reagent. Just before beginning the set of analyses, prepare a fresh oxidizing reagent by mixing 20.0 ml hypochlorite solution with 80 ml alkaline citrate solution.

Ammonium stock solution (1 ml = 1.00 mg NH₃-N). Dissolve 3.819 g anhydrous ammonium chloride (NH_4Cl) that has been dried for several hours at 100° C in demineralized water, and dilute to 1,000 ml with demineralized water.

Ammonium standard solution (1 ml = 0.010 mg NH₃-N). Dilute 1.00 ml ammonium stock solution to 100.0 ml with demineralized water.

Procedure

1. As soon as possible after sample collection (in the field, if possible) place 50.0 ml sample in the stoppered bottle or graduate.

2. Add 2.0 ml phenol reagent, stopper, and mix.

3. Add 2.0 ml nitroprusside catalyst reagent, stopper, and mix.

4. Add 5.0 ml freshly mixed oxidizing reagent, stopper, and mix.

5. Place out of direct light, and allow the blue color to develop for at least 1 hour. The color is stable for up to 24 hours, which allows one to return to the laboratory if steps 1 through 4 were done in the field.

6. Measure the absorbance using 1-inch test tubes at 640 nm. Determine the nitrogen concentration from a standard curve.

Standard curve for ammonia nitrogen concentration of the original water sample. For many natural waters, the dilutions shown in Table 6 have been found suitable.

Nitrate nitrogen

Nitrate nitrogen, although usually present in low concentrations in natural water, is often the most abundant inorganic form of the element. Natural concentrations rarely exceed 10 mg N per liter and are frequently less than 1 mg N per liter, especially during periods of high primary production. High concentrations of nitrate nitrogen (greater than 20 mg/L) may be a health hazard to juvenile mammals. Nitrates are reduced to nitrites in the digestive system. Nitrites are capable of oxidizing hemoglobin to produce methemoglobinemia. Hemoglobin in this form is incapable of transporting oxygen.

The phenoldisulfonic acid method (American Public Health Association, 1976) is sensitive, but suffers from severe interferences that must be removed by time-consuming techniques. The brucine method (American Public Health Association, 1976) is less sensitive and requires careful temperature control.

The cadmium reduction method is the method of choice for most natural waters with low concentrations of NO_3-N. When a water sample is slowly passed through a column containing amalgamated or copperized cadmium filings, or is vigorously shaken in a closed container with these filings, the nitrate is reduced to nitrite. The nitrite is then determined by the reliable and sensitive diazotization method. The column procedure is well described in *Standard Methods,* (American Public Health Association, 1976) which uses the mercury-cadmium packing and by Stainton, Capel, and Armstrong (1974) who use the copper-cadmium packing.

The following simplified cadmium reduction method, using prepared Hach reagents is often adequate, and if a mechanical shaker is used, has the advantage of allowing many samples to be reduced at one time. However because the reduction is dependent upon both time and vigor of shaking, identical treatment of samples and standards is essential. The time required for complete reduction tends to vary for different waters and for different concentrations of nitrate nitrogen. Therefore it is good practice to initially run internal spikes (known standards added to sample water) for increasing shaking times to determine efficiency of recovery. The cadmium reduction method measures both nitrate and nitrite nitrogen. Because nitrite is rarely measurable in unpolluted natural waters, a preliminary test to demonstrate its absence allows one to proceed with the cadmium reduction procedure, and to assume that all color development is due to nitrate nitrogen.

Apparatus

Spectronic 20 with blue phototube and set at 500 nm
Matched 1-inch test tubes
Glassware

40 to 50 ml glass bottles or flasks with stoppers.
Mechanical shaker (optional)

Table 7. Volume of standard solution (1.00 ml = 2.5 μg NO$_3$-N)
to be diluted to 100.0 ml and the resultant concentration

Milliliters of standard solution	Nitrate Nitrogen Concentration (μg/1)
1.0	25
2.0	50
4.0	100
8.0	200
16.0	400

Reagents

Hach NitraVer VI powder pillows
Hach Nitriver III powder pillows
Stock nitrate solution (1 ml = 100 μg NO$_3$-N). Dissolve 0.7218
g anhydrous potassium nitrate (KNO$_3$), and dilute to 1,000 ml with
demineralized water.
Standard nitrate solution (1.00 ml = 2.5 μg NO$_3$-N). Dilute 25
ml stock solution to 1000 ml with demineralized water. Prepare fresh
weekly.
*Standard curve for nitrate nitrogen concentration in the original
water sample.* For many natural waters, the dilutions shown in Table
7 have been found suitable.

Procedure

1. Add the contents on one NitriVer III powder pillow to a
25 ml sample in an Erlenmeyer flask. Shake for 30 seconds. If
a pink color develops within 10 minutes, nitrite nitrogen is
present. This may be quantified by starting with step 5 below.

2. Add the contents of one NitraVer VI powder pillow to a
30 ml water sample (or standard) in the glass bottle or flask
(NOTE: this is not the same sample as in 1 above). Stopper and
shake vigorously for at least 3 minutes. Be sure standards and
samples are shaken in exactly the same manner.

3. Wait for 30 seconds to allow the cadmium metal to settle.
Then decant 25 ml into a clean flask.

4. Add the contents of one NitriVer III powder pillow and
shake for 30 seconds.

5. If nitrate (or nitrite) nitrogen is present, a pink color will
develop. Allow the color to develop. After 10 minutes but before

20 minutes measure the absorbance using the 1-inch test tubes and the Spectronic 20 set at 500 nm. Determine the concentration of nitrogen from the standard curve. If nitrite was detected in step 1 but not quantified, report the results as combined nitrate and nitrite nitrogen. If nitrite was quantified, subtract that concentration from combined nitrate and nitrite concentration and report the difference as nitrate nitrogen.

SILICA

Silica is extremely common in nature as a constituent of igneous rocks, quartz, and sand. Natural waters often contain 1 to 10 mg silica per liter and rarely exceed 60 mg silica per liter. The chemistry of silica in solution is not known with certainty; however, most silica is present in the nonionized form. Silica is primarily of concern to the limnologist as the principal component of the cell wall of diatoms. A silica cycle occurs in bodies of water that contain diatoms. Silica is removed from the water during synthesis and slowly returned to the water by re-solution of the dead organisms.

This procedure measures only the molybdate-reactive silica. Certain forms of silica do not react with molybdate. The biological significance of these nonreactive forms is unknown. An optional pretreatment to convert any nonreactive silica to the reactive form is sometimes used and is described in *Standard Methods*. Since the following procedure omits this conversion step, results should be reported as molybdate-reactive silica.

The described procedure for silica may proceed in two stages: an initial stage that will measure silica concentrations greater than 2 mg/L and a second stage for concentrations less than 2 mg/L. With some experience, visual inspection of the intensity of the yellow color produced in stage one will guide the analyst in deciding if stage two is necessary. If the approximate concentration expected is unknown, two standard curves will be required.

Apparatus

Spectronic 20 with blue phototube, wavelength at 410 nm, or with red phototube, wavelength at 815 nm

1 cm matched test tubes
Pyrex glassware

Reagents

Distilled water. Use deionized water stored in polyethylene bottles for all reagents.

Hydrochloric acid. 1 + 1 v/v.

Ammonium molybdate reagent. Dissolve 5 g $(NH_4)_6Mo_7O_{24}$ · $4H_2O$ in distilled water with slow warming; then dilute to 50 ml. Filter if any insoluble material persists. Adjust the pH of this solution to between 7.0 and 8.0 with silica-free ammonium or sodium hydroxide, and store in a polyethylene bottle.

Oxalic acid solution. Dissolve 5 g $H_2C_2O_4$ · $2H_2O$ in distilled water, and dilute to 50 ml.

Silica standard solution. Silica stock solutions are difficult to prepare and standardize. It is recommended the commercially prepared standard solutions be purchased from a reputable supplier. A practical concentration is 1 ml = 0.01 mg silica.

Reducing agent for stage two. Dissolve 0.5 g 1-amino-2-naphthol-4-sulfonic acid and 1 g anhydrous sodium sulfite (Na_2SO_3) in 50 ml distilled water (warm, if necessary). Add this to a solution of 30 g sodium bisulfite ($NaHSO_3$) dissolved in 150 ml distilled water. Mix and filter into a polyethylene bottle. Store this reducing agent in the refrigerator. When the reagent becomes dark colored, it is no longer usable and must be discarded.

Procedure

Samples are to be stored in polyethylene bottles and may be kept under refrigeration for up to a week. Excessive silica concentrations are reported when samples are held for a period of weeks prior to analysis, during which time diatom populations decompose. In addition to possible contamination by glassware, other sources of error are the presence in the water of phosphate, turbidity, or both. Both phosphorus and silica react with the molybdate to form acids. However, the addition of oxalic acid destroys the molybdophosphoric acid, leaving only the molybdosilicic acid, which produces a yellow color proportional in intensity to the concentration of the silica. Turbidity, if obvious, must be removed by filtration through any filtration apparatus shown to be free of silica. The reagent blank must also be run through the same procedures, including filtration, so that it will detect and compensate for any contamination.

It is possible to compensate photometrically for turbidity, thus canceling its effect. This technique may be used in the silica procedure. Compensation is accomplished as follows: two identical aliquots of each sample are prepared. All reagents are added to one aliquot as described later; all reagents *except* the molybdate are added to the other aliquot in normal sequence. The aliquot without the molybdate is used to set the photometer to zero absorbance; then the corresponding aliquot with the molybdate is placed in the light path, and absorbance is read. This automatically corrects for turbidity present. Even if compensation is used, a distilled-water reagent blank must also be run to check for contamination. Filtration or compensation is not necessary unless the sample is obviously turbid.

Stage one

1. To a 50 ml sample, add in quick succession 1.0 ml 1 + 1 HCl and 2.0 ml ammonium molybdate reagent, and mix by swirling.

2. Allow treated sample to stand for 5 to 10 minutes.

3. Add 1.5 ml oxalic acid solution, and mix well.

4. After 2 minutes, but before 15 minutes, measure the absorbance of the yellow color at 410 nm using 1 cm test tubes.

Stage two. If little yellow color develops, the silica concentration is low, and the reduction step to produce a blue color is necessary.

5. Within 15 minutes of the addition of the oxalic acid (step 3), add 2.0 ml of reducing agent and mix. Read the blue color developed after 5 minutes, using 1 cm matched tubes, with Spectronic 20 set at 815 nm.

6. Determine mg SiO_2 from a standard curve pre-

Table 8. Volume of standard solution (1.00 ml = 0.010 mg silica) to be diluted to 50.0 ml, with the resultant concentration

Milliliters of standard solution	Silica concentration (mg/liter)
For stage one determinations	
10.0	2.0
20.0	4.0
30.0	6.0
40.0	8.0
50.0 (no dilution)	10.0
For stage two determinations	
1.0	0.2
3.0	0.6
5.0	1.0
7.0	1.4
10.0	2.0

pared by carrying standards through either stage one or stage two from dilutions of silica standard solution.

Standard curve for the silica concentration in the original water sample. For many natural waters, the dilutions shown in Table 8 have been found suitable.

SULFATE

In many natural waters, sulfate is the second most common anion, being derived from most sedimentary rocks. Sulfate concentrations in natural waters are usually from 5 to 50 mg/L; however, much greater concentrations are found in saline lakes. Drinking water with a sulfate concentration greater than 250 mg SO_4 per liter may have a cathartic action. In lakes, sulfur is cyclic and involves organically reduced forms as well as the common free $SO_4^=$ ion. Although few studies have been conducted on sulfur metabolism in algae, sulfate is the common form taken up by higher plants and is probably equally important to algae. The turbidimetric method described here is one of the least precise, but because of its simplicity, it is often used for routine work. As described, the method is basically that of *Standard Methods* but makes use of the convenient Hach reagents.

Apparatus

Spectronic 20 with blue phototube set at 420 nm
Matched 1-inch test tubes
Glassware

Reagents

Standard sulfate solution (1.00 ml = 0.10 mg SO₄). Using a microburette, measure 10.41 ml standard 0.02 N H_2SO_4 titrant (from alkalinity procedure) into a 100 ml volumetric flask, and dilute to mark with distilled water.

SulfaVer powder. From Hach Chemical Co.

Procedure

Carry a distilled-water blank through all steps.

1. To a 25 ml sample in a 125 ml flask add 1.0 g SulfaVer powder, and swirl evenly for 1 minute. A suspension of barium sulfate forms.

2. Pour entire sample into a 1-inch matched test tube, and let stand for 3 minutes.

3. Read absorbance produced by this suspended turbidity at a wavelength of 420 nm. Estimate milligrams of sulfate by comparing with a standard curve prepared by running known standard concentrations through the

Table 9. Volume of standard solution (1.00 ml = 0.10 mg sulfate) to be diluted to 25.0 ml, with the resultant concentration

Milliliters of standard solution	Sulfate concentration (mg/L)
2.0	8.0
4.0	16.0
6.0	24.0
8.0	32.0
10.0	40.0

same procedure, *being careful that all timing is the same*. The highest standard should not exceed 40 mg/L (1 mg/25 ml sample), since this method fails above that concentration.

Standard curve for the sulfate concentration in the original water sample. For many natural waters, the dilutions shown in Table 9 have been found suitable.

SPECIFIC CONDUCTANCE

Specific conductance is a measure of a water's capacity to conduct an electric current. It is the reciprocal of resistance for which the standard unit is an ohm. Since conductance is the inverse of resistance, the unit of conductance is the *mho*, or in low-conductivity natural waters, the *micromho*. Because the measurement is made using two electrodes placed 1 cm apart, specific conductance is generally reported as micromhos per centimeter. The relationship of specific conductance to ionized matter concentration varies with both the quality and quantity of the ions present. However, at low concentrations the ions move and behave independently, and the conductance-concentration relationship is almost linear. Temperature of the solution affects the ionic velocity and, thus, the specific conductance. In unpolluted waters, conductance increases from 2% to 3% per degree Celsius; generally a correction of 2.5% is usable. Since temperature is an integral part of these data, it must always be reported along with the specific conductance. For convenience, most literature values are found corrected to one standard temperature. Sanitary engineers and geochemists have long used 25° C as a standard temperature. Recently limnologists have used 18° C as standard, since this is closer to the average natural water temperature found in the United States.

The greatest value of specific conductance to the limnologist is the estimation, by this simple procedure, of the total concentration of dissolved ionic matter in the water, which in turn is related to water fertility. Total dissolved matter may be estimated from conductance data by multiplying by some empirical factor, which usually varies from 0.5 to 1.0 (This method is considered under the procedure for dissolved matter.) Specific conductance values also provide a quick check of alteration of the total quality of water because of the addition of many pollutants. With experience, the analyst may also use conductance data to estimate the aliquots of samples to be taken for many chemical analyses so that concentrations will fall within the range of the method.

Apparatus

Conductivity meter with dip cell (Fig. 13)
Water bath at 18° or 25° C, if conductivity meter lacks temperature compensation

Fig. 13. Materials needed to determine specific conductance include an electric bridge (meter) and dip cell, one container for the water sample and a second for distilled water, and a thermometer.

Thermometer
2 beakers of 200 to 250 ml capacity or similar vessel deep
    enough to cover vertical dip cell

Procedure

1. Connect lugs of dip cell to instrument (the lugs may be connected to either post). A cell constant of 1 is best for most natural waters; other cells or ranges are available for weak or strong conductors. If the cell constant is unknown, or if it is suspected that through time the constant has changed, the actual cell constant may be calculated as follows: prepare a 0.00702 N potassium chloride solution. (Dissolve 0.5232 g KCl, oven dried at 180° C for 1 hour, in demineralized water, and dilute to 1,000 ml.) This electrolyte has a specific conductance of 0.001000 mho at 25 ° C. The cell constant is calculated by multiplying the measured resistance of this electrolyte solution at 25° C by the known specific conductance.

2. Fill two beakers with water sample to a depth sufficient to cover the vent holes in the dip cell when immersed. Place in water bath, and wait for equilibrium if necessary.

3. Measure temperature of the second beaker, and set temperature compensator on instrument to that temperature.

4. Rinse dip cell thoroughly in the first beaker, and then transfer dip cell to second beaker. Turn on instrument, and, while gently moving the dip cell, keeping the holes submerged all the time, take a reading of specific conductance.

5. Turn off instrument, remove and disconnect dip cell, rinse with distilled water, and air dry.

RESIDUE

Residue refers to material left in a vessel after evaporation of a water sample. It may be subdivided into organic matter and inorganic material. These may both be further subdivided into particulate and dissolved matter.

Since variables are numerous in this type of determination, the accuracy is usually not as great as for most limnological determinations. Three procedures are described next. Total residue (sometimes referred to as suspended matter) is simply the weight of material remaining after evaporation of raw water. Filterable residue (dissolved matter or dissolved solids) is the weight of material remaining after evaporation of the filtrate of the raw water sample. Volatile matter (conversely fixed residue) is the difference between the weight obtained in either of the other procedures (specified) and that obtained after ignition at 600° C.

Total residue
Apparatus

Evaporating dishes
Drying oven at 103° C
Analytical balance

Procedure

Collect sample in Pyrex or polyethylene bottle.

1. The dish to be used to hold the sample must be pretreated in the same manner as used on the sample (Fig. 14). This means that if the sample is to be dried in an oven and then placed in a desiccator before weighing, the empty dish must be taken through the same series of steps to determine the weight of the empty dish. If ignition is to be used, the empty dish must be ignited, cooled in a desiccator, and weighed for initial weight.

2. To the prepared and preweighed dish add a well-mixed aliquot of water sample to be evaporated. Usually 100 ml will suffice, but exact volume will be determined by the sensitivity of the balance. It must not contain more than 200 mg of residue.

3. Evaporate to dryness, and then dry overnight in an oven at 103° C.

4. Remove dish from oven, cool briefly in the air, and

Fig. 14. The volatile matter portion of total residue is determined by igniting the evaporated and weighed sample in a muffle furnace at 600° C. The crucible containing the ash is cooled in a desiccator and reweighed. The difference in weights is volatile matter.

place in desiccator; dishes must not touch each other or the sides of the desiccator.

5. Take desiccator to balance, remove cooled dishes, and weigh rapidly.

Calculations

$$\text{mg total residue on drying at } 103° \text{ C per liter} = \frac{\text{mg total residue} \times 1{,}000}{\text{ml of sample}}$$

Filterable residue (dissolved solids)

Apparatus

As for total residue, plus a filter apparatus using acid-washed, ashless, hard-finish filter paper for fine precipitates, or for most precise work, a filter apparatus using 0.45 μm membrane filters

Procedure

The procedure is the same as for total residue except that the sample is filtered and the filtrate evaporated. Results are reported as filterable residue or total dissolved solids on drying at 103° C.

Since, as pointed out in the section on specific conductance, the concentration of dissolved ionic material may be generally related to the conductivity of the water, it is possible to estimate total dissolved matter from specific conductance data. This is accomplished by multiplying the specific conductance by some empirical factor that is usually less than 1. Rainwater and Thatcher (1960) suggest that 0.65 times specific conductance approximates filterable residue (total dissolved matter or solids) in most natural waters. However, it is best to determine this factor for each body of water routinely studied. This is done simply by determining the specific conductance at a stated temperature of the same water to be used for determination of filterable residue. The factor (f) is then determined by dividing the filterable residue by specific conductance. Once this is determined, one may avoid using the evaporation method on a regular schedule and employ it only to occasionally check the validity of the f factor.

Volatile matter

Volatile matter provides only a crude estimate of organic matter present in the water sample mainly because ignition also produces a decomposition or volatilization of inorganic salts. Better estimates of organic matter, especially in low concentrations, may be obtained by quantitative dichromate oxidation (Maciolek, 1962) or infrared carbon analysis. However, the volatile matter procedure is more simple because the sample used in determination of total or filterable residue is carried on through only one more step.

Apparatus

As for residue, but vessel must be able to withstand higher temperatures and must be preignited before initial weight is determined; also, an electric muffle furnace is required.

Procedure

1. Preheat muffle furnace to 600° C.
2. Ignite sample used in residue determination for 15 minutes, air cool briefly, complete cooling in desiccator, and weigh. The loss on ignition is reported as milligrams of volatile solids per liter.

Calculations

$$\text{mg volatile solids per liter} = \frac{\text{mg volatile solids} \times 1,000}{\text{ml of sample}}$$

3 □ Plankton

Plankton are organisms that, because of their size, immobility, or both, are at the mercy of water movements. The limnologist generally considers these to be tiny forms of life. However, this group encompasses organisms whose sizes span three orders of magnitude. Plankton include forms of aquatic bacteria and ultra-algae only a few microns in diameter and macroscopic forms of crustacea several millimeters long. Thus the plankton encompass a wider range of sizes than do the larger, macroscopic nekton.

Most plankton sampling techniques are size selective. A convenient, if artificial, size differentiation of the plankton is net plankton and nannoplankton. The distinction is based on the passage of organisms through a no. 25 mesh plankton net (Table 10). The larger organisms retained by the net, that is, "net plankton," tend to be predominately zooplankton consisting of cladocerans, copepods, and large rotifers, but also include large or colonial phytoplankton. The smaller organisms that pass through these meshes (ca 64 μm), "nannoplankton," are predominately phytoplankton and bacterioplankton, although protozoa and small rotifers are frequently present.

Plankton are important components of aquatic systems. In larger lakes and reservoirs the plankton are the major primary and secondary links in the trophic rela-

100

Table 10. Plankton net bolting cloth specifications

Silk size no.	Meshes per inch	Silk aperture (μ)	Nylon aperture (μ)
0000	16	1,364	1,340
000	23	1,024	1,050
00	29	752	752
0	38	569	571
1	48	417	423
2	54	366	363
3	58	333	333
4	62	318	316
5	66	282	280
6	74	239	243
7	82	224	223
8	86	203	202
9	97	168	165
10	109	158	153
11	116	145	—
12	125	119	116
13	129	112	110
14	139	99	102
15	150	94	93
16	157	86	86
17	163	81	—
18	166	79	—
19	169	77	—
20	173	76	80
21	178	69	—
25	200	64	64

tionship. This significance decreases as the volume of water decreases; in smaller, shallower waters the benthic forms become increasingly important.

Although large rivers may have a true plankton community, the potamoplankton, most plankton of rivers are a transient community (tachyplankton), having been washed in from lake, pond, and reservoir drainages. Plankton organisms are interesting to the biologist because of their physical nature. The ability to maintain position in an open water system requires that most of these organisms have unusual modifications for flotation. Many groups, although lacking obvious means of motility, have regular diel vertical movements. The mechanisms and stimuli for such movements have interested aquatic biologists for some time. Also, the plankton often serve as indicators of water quality. Chemical conditions play a part in determining the taxonomic nature of the plankton, and the species composition is often used not only to classify the water as polluted or free from pollution but also to suggest the quantities of various naturally occurring substances, such as nitrogen and phosphorus.

In sampling the plankton, one must first determine what group of organisms are desired and from what portion of the aquatic system they are to come. Several devices for collecting these organisms have been devised. Three "net" methods and one "nanno" method are described next. A net towed behind a boat for some distance provides the greatest quantity of plankton with minimum effort. An advanced design of this type is the Clarke-Bumpus net, which is a net with a metering device to determine the quantity of water passing through it (Fig. 15). Although the net is excellent in many respects, its major weakness is that precise depths are difficult to sample because a slight change in the movement of the boat causes the depth of the net to vary.

A second method employs the Wisconsin net (this net can likewise be used as a tow net), in which known quantities of water are collected from discrete depths by the use of a water sampler or pump and are poured through the net (Fig. 17). This method assures collection from precise depths but has a serious drawback in that the size of the sample is relatively small because of the time and effort involved in collecting many samples with the usual 1- to 3-liter water-sampling bottles. A plankton pump that empties through the net overcomes the volume problem but introduces additional mechanical problems in the field, as well as possible avoidance by larger zooplankton.

A third method is the use of a "trap," which is a device that combines a discrete depth sampler with a net (Figs. 18 and 19). The volume of a trap is typically much greater than that of a water sampler, thus reducing the number of "hauls" required for an adequate sample. The trap is a box that is lowered to the desired depth with the top and bottom open. The openings are closed to encompass a known volume of water, which drains through a plankton net when the trap is lifted from the water. The principal advantage of this sampler is the ability to sample precise depths; the fact that a sample is both collected and concentrated in one step makes this type of sampler a favorite of many investigators of the net plankton.

Nannoplankton are organisms that pass through a no. 25 net. To collect these organisms, it is obvious that water must also be passed through such a net and be collected. The tiny organisms must later be separated from the water by some other method, such as settling, filtration, or centrifugation, all of which are done in the laboratory. The usual field technique is to collect the water that was passed through the Wisconsin net as described before and preserve it for further nannoplankton analysis.

The following procedures describe methods of collecting plankton from discrete depths. Although collection from separate depths is usually the preferred technique for quantitative studies, it is possible to entirely miss some plankton organisms because of their presence in a thin "plate" located between the preselected sampling depths. A vertical tow started at a selected depth or at the bottom will sample plankton at all depths. Although vertical tows are primarily used for qualitative work, quantitative data may be obtained by calculating the volume of water sampled as follows: Volume sampled = Length of tow $\times \pi \times$ (Radius of net opening)^2. If the length of vertical tow is insufficient to give an adequate concentration of organisms, an oblique tow should be made, from a preselected depth or from the bottom, behind a moving boat. It is important that the net be retrieved to the surface slowly and at a regular rate to avoid bias of oversampling or undersampling at any one depth.

CLARKE-BUMPUS NET
Apparatus

Clarke-Bumpus net with 2 messengers, buckets, and calibrated line (Fig. 15)
100 to 200 ml wide-mouthed bottles or baby food jars
Plastic "squeeze" bottle
Clinometer

Reagents

Neutral formalin (10%). Saturate 37% formaldehyde solution with magnesium carbonate, then dilute one part with nine parts water. NOTE: Some add a few drops of liquid detergent per liter. This reduces surface tension and prevents "clumping" of the organisms. NOTE: Some, especially those working with Cladocera and Rotifera, prefer an "osmotically balanced" formalin solution to prevent distortion. This solution is prepared by adding 80 g/L of sucrose to the 10% formalin solution, which when diluted in use produces a final sucrose concentration of approximately 0.12 molar (Haney and Hall, 1973).

Carbonated water. To prevent contraction and distortion from the shock of preservation, zooplankton may be anesthetized before preservation by immersing the bucket of the plankton net in carbonated water for approximately 1 minute. Bottled carbonated water may be carried to the lake or prepared on site by adding one Alka-Seltzer tablet to 200 to 300 ml water in a container of sufficient size to hold the plankton net bucket.

Procedure

1. Select portion of lake to be sampled, assemble Clarke-Bumpus net, using net of desired mesh for the type of organisms wanted. Be sure that net and bucket mesh are of the same size. Attach free end of calibrated line to boat.

2. Close door mechanism of net, take and record meter readings, and lower net over the side of the boat.

Fig. 15. Clarke-Bumpus metering plankton net, shown with the door in the intermediate, or open, position. When the net is being towed, the lateral planes cause the net to ride in a position approximately at a right angle to the metal suspending frame.

The depth to be sampled may be determined by the use of the trigonometric function for a right triangle. The line trailing behind the boat forms the hypotenuse of this triangle. Using a clinometer (one can easily be made from a protractor), determine depth of tow by multiplying length of hypotenuse by the cosine of the angle as measured by the clinometer, or read from nomograph (Fig. 16).

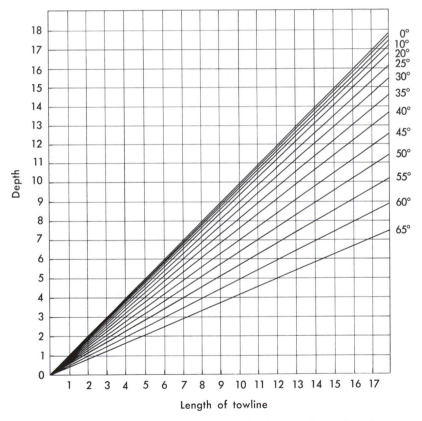

Fig. 16. Nomograph for determining depth of a towed sampler when length of towline and angle of tow are known.

3. Once the boat speed is maintained and steadied and the desired depth is reached, drop the first messenger down the line to open the net. Tow for a measured period of time (try 1 minute to start), and then drop the second messenger to close the net.

4. Retrieve net and wash up and down while lifting the net from the water to gradually wash organisms from the sides of net down into the bucket. Take meter reading. The number of revolutions turned while the net is opened is proportional to the volume of water passing through the net. This relationship of volume to revolutions is usually determined by the manufacturer and recorded for that particular net; however, with time, corrosion, and wear, this calibration changes, and it is important to recalibrate the net's meter by moving it through a known distance in the water and calculating the volume passed through per revolution. Since you recorded the time of the sampling run for a given boat speed, you now have an approximation of how long a net must be in the water at that speed to filter a desired quantity of water. This is useful for future tows.

5. Drain the excess water through the screen of the bucket so that the plankton are contained in 25 to 50 ml water. Disassemble the bucket, and pour this water with plankton into a jar. Use the plastic squeeze bottle with 10% formalin to wash the remaining organisms from the bucket. Continue this washing until all organisms have been washed into the bottle. Continue to add 10% formalin until the volume of the original water in the jar is doubled. This will give the desired final 4% to 5% concentration for preservation. If excess 10% formalin is added, dilute down to 4% to 5% using filtered lake water. Label the bottle with data for the collection.

6. Wash out net thoroughly, and reassemble for next tow. It is important that all plankton nets always be well washed at the end of a day's sampling and again with mild detergent when the nets are returned to the laboratory. This prevents gradual plugging of the nets and possible carry-over of organisms from one sample to another.

Fig. 17. The popular Wisconsin plankton net. It may be used as a tow net, either vertically, obliquely, or horizontally, or as a filter net through which water is poured. (Courtesy Wildlife Supply Company, Saginaw, Mich. 48602.)

WISCONSIN NET FOR NET PLANKTON AND NANNOPLANKTON

Apparatus

No. 20 or 25 Wisconsin plankton net (Fig. 17)
Plastic pail
Kemmerer or van Dorn water-sampling bottle
100 to 200 ml wide-mouthed bottles or baby food jars
1- to 2-liter polyethylene bottles

Reagents

Supply of 10% neutral formalin with sugar (p. 104) or supply of 95% alcohol (ethyl or isopropyl).

Lugol's iodine. Dissolve 5 g iodine (I), 10 g potassium iodide (KI), and 10 ml glacial acetic acid in 100 ml demineralized or distilled water. *Or* supply of concentrated (100%) neutral formalin with sugar (p. 104).

Procedure

1. Select a sampling site, and anchor boat to prevent drift. Choose sampling depths based on physical, chemical, or morphological features of the lake.

2. Assemble Wisconsin net with bucket, and collect water samples from the desired depth. These samples are strained through the mouth of the Wisconsin net, and the water is collected in the plastic bucket. Be sure no water splashes over the outside edge of the Wisconsin net. Count and record (it is easy to forget) number of water samplers emptied through the net. Determine volume of water filtered by multiplying the number of samplers emptied by the volume of the sampler used. Although the actual volume necessary to provide a concentrate adequate for most studies will vary widely depending on the nature of the water, a minimum of 30 liters is usually required in eutrophic waters and much more in oligotrophic waters.

3. Wash down organisms inside plankton net by sloshing it over side of boat, but not allowing water to flow over mouth of net, while gradually raising net from the water so that organisms are concentrated in the buck-

et. Disassemble bucket, and while holding it over the mouth of the jar, remove stopper and allow contents (25 to 50 ml) to drain into jar. Wash down sides of bucket with 10% formalin from plastic squeeze bottle to double original water volume in jar. This will produce the desired 4% to 5% concentration for preservation. Label with collection data.

4. Pour the material that passed through the net but was collected in the plastic bucket into the large plastic bottle and preserve with Lugol's iodine. Sufficient Lugol's iodine is added to the water to make a 1% solution (color of weak tea). A 4% to 5% formalin solution may be used if Lugol's iodine is unavailable.

5. Label with collection data, and store Lugol's preserved sample in a dark place.

TRAPS FOR ZOOPLANKTON

The Juday plankton trap has been used for several decades, whereas the Schindler-Patalas plankton trap has recently come into use (Schindler, 1969) (Figs. 18 and 19). The Juday trap has the disadvantage of smaller volume and, although made of metal, is subject to damage, especially to the slide guides of the door-closing mechanisms. The Schindler-Patalas trap also requires great care in use because it is constructed of clear Lucite or Plexiglas that may fracture if struck sharply, as on the side of a boat. A definite advantage of the Schindler-Patalas trap, in addition to its volume and transparency (possible plankton avoidance problems), is that it may be constructed in a shop with only simple tools (it is also commercially available).

Juday trap

Apparatus

Juday trap (Fig. 18), messenger, plankton bucket of appropriate mesh, and calibrated line
100 to 200 ml wide-mouthed bottles or baby food jars
Plastic squeeze bottle

Fig. 18. The Juday plankton trap is one of the best methods for sampling at discrete depths. When snapped shut, the top and bottom plates slide over the sides, enclosing a volume of water that drains through the net of the bucket when the trap is lifted from the water. (Courtesy Wildlife Supply Company, Saginaw, Mich. 48602.)

Fig. 19. Student-constructed Schindler-Patalas plankton trap. The upper and lower doors are shown partially open. The plankton bucket, salvaged from an old Wisconsin net, may be replaced by a small plankton net to permit more rapid drainage of the trap.

Reagents

Supply of 10% neutral formalin with sugar (p. 104) or supply of 95% alcohol (ethyl or isopropyl)

Procedure

1. Attach line to trap by passing it through hole in trip lever and then securing it beneath hole in plate attached to the four suspending chains (if a solid messenger is used, slip on line before attaching trap). Inspect grooves of sliding door mechanism for cleanliness, and check freedom of movement of doors. Attach plankton bucket. Open doors, and set trip lever mechanism's knife edge behind block (weight of trap must be kept on line from this point).

2. Lower trap to desired depth. Small traps up to 5-liter capacity may be lowered by hand, but larger traps require a winch and boom. As the trap is lowered, a column of water is cut so that at any instant the contained volume represents organisms from that depth.

3. Drop messenger to close trap. Although avoidance by organisms is minimal as the trap descends, larger and more motile forms may escape from the trap if there is unnecessary delay in closing. With practice, one can drop the messenger slightly before trap reaches desired depth so that closure and time of reaching that depth are almost simultaneous.

4. Retrieve trap, and remove the bucket. Drain plankton and a small amount of water (25 to 50 ml) into jar. Wash down the sides of the bucket with 10% formalin from the plastic squeeze bottle. Add sufficient 10% formalin to double the original volume in the jar. This will produce the final desired 4% to 5% concentration for preservation. NOTE: Some prefer to preserve zooplankton in 70% ethyl or isopropyl alcohol because microdissection and manipulation is easier on the more flexible specimens. If this is desired, use a smaller original volume of water with the plankton in the jar, and use 95% alcohol for washing the bucket.

5. For most water, a single 5-liter sample is inadequate to assure representation of most species or quantitative sampling accuracy. Consequently, replicates are taken from the same site and depth. These may be pooled in a single jar for survey work or kept separate when statistical analysis is desired. Often, 20 liters is an adequate volume for many mesotrophic to eutrophic lakes and reservoirs.

6. Label the jar with the appropriate collection data, including number of trap hauls composited into one sample.

7. Before storing trap, inspect grooves for cleanliness. It is also good practice to spray the slides, grooves, and springs with an aerosol water-dispersing lubricant such as WD-40.

Schindler-Patalas trap

Apparatus

Schindler-Patalas trap (Fig. 19) with net and/or bucket of appropriate mesh and calibrated line
100 to 200 ml wide-mouthed bottles or baby food jars
Plastic squeeze bottle

Reagents

Supply of 10% neutral formalin (p. 104) or supply of 95% alcohol (ethyl or isopropyl)

Procedure

1. Attach line to trap so that depths are measured from the midpoint of the box.

2. Attach appropriate mesh net and/or bucket.

3. Check to make sure that upper and lower doors are moving freely.

4. Lower smoothly to the desired depth, and immediately begin a smooth and continuous retrieval. Do not allow any downward motion, since this could allow some opening of the lower door even though the upper door is now locked in place by a snap-lock mechanism. If

used when there is surface wave action, it is especially important to lift the trap quickly above the surface of the water. Be careful to prevent trap from striking edge of boat because it is rather easily broken.

5. While holding over the side of the boat and well above any wave, allow all contained water to drain through the plankton net and/or bucket. Wash down net so as to wash plankton into bucket.

6. Dissemble bucket, and drain contents along with a small amount (25 to 50 ml) of water into the jar. Wash sides of bucket with 10% formalin so that the original volume of water in the bottle is doubled. This will produce the 4% to 5% concentration desired for preservation. NOTE: Some prefer to preserve zooplankton in 70% ethyl or isopropyl alcohol, since microdissection and manipulation are easier on the more flexible specimens. If this is desired, use a smaller original volume of water with the plankton in the jar, and use 95% alcohol in the plastic squeeze bottle for washing the bucket.

NANNOPLANKTON LABORATORY METHODS

Any method for the identification and enumeration of nannoplankton must permit examination of these materials under a compound microscope, using a minimum of the high-power dry objective (430×) and preferably using oil immersion (1,000×). Many of these organisms are delicate, and preparation for examination requires attention to avoid fracturing, rupturing, or plasmolyzing them. Most methods of preparing nannoplankton for identification and enumeration have some undesirable features. It is usually necessary to concentrate the sample before statistically reliable counts are obtained. Centrifugation and filtration each will result in some damage to the cells, although this may be lessened by using slow centrifuge speeds for longer periods and by using low filtration vacuum. Sedimentation offers a gentle method with little cell damage, but it is very slow and limits the number of samples that may be processed. The membrane filter method, which is one of the more widely used procedures, and the sedimentation tube method using an inverted microscope are described next. Other methods that the analyst may wish to consider include the Palmer cell and the direct drop count. Either of these may be used with raw preserved water or with concentrates from centrifugation or from sedimentation tubes after siphoning off the superficial water. The Palmer cell is a counting chamber on the same principle as the Sedgwick-Rafter cell (pp. 126 to 128) but is only 0.4 mm deep, permitting use of higher magnification microscope objectives. The drop count method, as developed by W. T. Edmondson and described in Vollenweider (1969), offers simplicity.

Although identification of most phytoplankton taxa is easiest using fresh, or at least undamaged, preserved material, the identification of diatoms, a major component of many phytoplankton and periphyton communities, is best done with specimens that have been cleaned of their cytoplasmic components. A description of a procedure for cleaning diatoms and their subsequent mounting for examination is included.

Membrane filter method

Apparatus

  1-inch diameter gridded membrane filters, pore size 0.8 μm
  Vacuum filtration apparatus
  Graduated cylinder
  Clearing oil (immersion or cedarwood oil)

Procedure

1. Measure out a portion of the well-shaken, preserved nannoplankton sample into a graduated cylinder. In many waters 25 ml may be appropriate.

2. Filter this quantity through the membrane filter, *using very low vacuum*. Add contents of graduated cylinder all at once to promote a random distribution of organisms over the surface of the membrane filter.

3. Remove filter from apparatus, handling by the

edges, and place filter on cleaned microscope slide. Align horizontal lines of grid parallel to edge of microscope slide.

4. Place a drop of clearing oil on the microscope slide at each edge of membrane filter. *Do not* cover filter with oil. The center portion of the filter must remain oil free to permit evaporation of water for drying of the slide. As the slide dries, the oil from the edges will replace the water (Fig. 20) from the edges. When dried, the membrane filter will be transparent. A milky filter coloration indicates that the filter became oil coated without complete drying. The time for drying and clearing depends on the air moisture conditions but is usually complete within 2 to 4 hours. During this drying, the slide should be stored in a dust-free place.

5. The microscope slide is now ready for examination with or without a coverslip. One may work with the oil immersion lens directly on the membrane filter if an additional drop of oil is added. With this preparation, most organisms may be both identified and counted on the same slide. Common forms of life observed in the nannoplankton will be phytoplankton, rotifers, and some protozoa.

6. Organisms may be counted by the total count, strip count, or field count method, depending on the numbers of organisms present. A simple method to estimate when sufficient fields have been counted is the changing average method. To do this, count from an initial selected set of fields (for example, 10 fields). Analyze these counts, and select an organism that appears regularly but not in high numbers. Calculate the average number of these organisms per field. Then count an additional field, add the number of organisms, and recalculate the average. If the average was not changed by the additional field, then sufficient fields have been counted; however, if the average was changed by the additional field, continue to count additional fields until no further change in the average frequency of the organism is determined. A tabulation sheet such as that shown in Fig. 21 is helpful.

Fig. 20. Gridded membrane filters for quantitative nannoplankton analysis. An aliquot of lake water is gently filtered and the damp filter immediately placed on a clean microscope slide *(top)*. Several drops of oil are placed on the slide at the edge of the filter. As the filter dries, oil is drawn into the filter matrix, rendering it transparent *(center)*. The completely oil-impregnated clear filter is now ready for microscopic examination *(bottom)*.

Calculations

Number of organisms of each species per aliquot filtered =

$$\frac{\text{Number of organisms in fields counted}}{\text{Fields counted}} \times \text{Number of fields in filtered area}$$

Number of organisms of each species per liter of lake water =

$$\frac{(\text{Number of organisms in aliquot filtered}) \times 1,000}{\text{Volume filtered (ml)}}$$

Fig. 21. Use of a standard form for enumeration of plankton is convenient and assures that all necessary data are permanently recorded. Identified organisms are listed in the left-hand column. Space is provided to record the number present in each of ten different field or strip counts.

PREPARATION OF PERMANENT MEMBRANE-FILTERED NANNOPLANKTON SLIDES

The membrane filter method using immersion or cedarwood oil provides relatively permanent slides if kept in a dust-free place. Cedarwood oil tends to evaporate more rapidly than immersion oil and must be replaced occasionally, but both will provide slides that will last for months. If permanent slides are desired, the following procedure may be used.

Reagents

Glutaraldehyde-ethanol solution. Add 5 drops of glutaraldehyde, with mixing, to 50 ml distilled water. Refrigerate for 6 hours. Then mix with 50 ml ethyl alcohol.

Beechwood creosote–ethanol mixture. Mix 7 ml 95% ethyl alcohol with 3 ml beechwood creosote.

Procedure

1. *Gently* filter sample as before, but do not filter dry. Leave approximately 5 ml water in the filter apparatus, and add 10 to 15 ml of the glutaraldehyde-ethanol mixture without mixing. Filter down to 5 ml.

2. Add 10 ml of the beechwood creosote–ethanol mixture, and filter until only a very thin liquid film remains on the filter.

3. Remove filter, and place on microscope slide. Saturate with pure beechwood creosote. Add coverslip, and set aside for approximately 2 weeks to clear.

4. Clean filtration apparatus with 95% ethanol.

Sedimentation tube–inverted microscope method

The sedimentation tube–inverted microscope method has the distinct advantage of being the most gentle for soft-bodied forms. It provides the most transparent slides and, therefore, is best for identification requiring work with minute structures. The major disadvantages are the time required for the sedimentation process to occur and the necessity of an inverted microscope. For all but advanced studies, it is doubtful that the advantages of this method would offset the disadvantages when compared with the membrane filter technique.

Apparatus

Inverted microscope

Sedimentation cylinders, which may be purchased or made in a variety of sizes. Typical size ranges are from 5 to 25 ml ca-

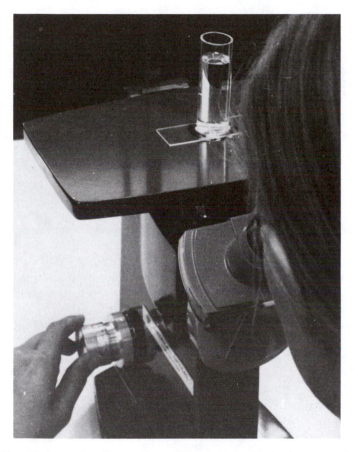

Fig. 22. Use of sedimentation tube on an inverted microscope is especially valuable in the analysis of delicate plankton organisms that may be unrecognizably damaged by other techniques of concentration.

pacity. To construct these in a laboratory, one may use either glass or Lucite tubing 22 to 25 mm in diameter (Fig. 22). A microscope slide or a large rectangular or circular glass coverslip is cemented to one end with a minimum of cement. To assure complete sedimentation of small forms, the height of the contained water column must not exceed five times the diameter of the cylinder. A small, flat square of Lucite or glass should also be available to serve as a temporary cover for the cylinder.

Procedure

1. Measure out a known volume of a well-shaken nannoplankton sample that has been preserved in Lugol's iodine, and pour into the cylinder.

2. Cover and set cylinder aside in a vibration-free place near the inverted microscope to allow sedimentation to occur. The time required for sedimentation of small organisms can be calculated by allowing 4 hours for each centimeter of water column.

3. Following sedimentation, place cylinder on an inverted microscope stage, and take either total, field, or strip counts. The magnification under which these counts are made will vary with the size of the organism encountered. Make a cursory examination of the cylinder under oil immersion to determine if there are very small forms present. If so, identification and enumeration should be at this magnification. A total count in even moderately rich waters under this magnification would likely require many hours. In this case, count random fields, using a Whipple ocular disk that has been calibrated by stage micrometer, or, using opposite parallel edges of the Whipple disk, count strips across the field until adequate numbers are obtained. Count until there is no further change in the average occurrence per field as indicated before or, as some investigators prefer, until you have counted a hundred organisms of average frequency.

Preparation of diatoms for examination

Diatoms are encased in a silica wall or frustrule. The surface features of the frustrule provide the main characteristics used in identification of the diatoms. It is necessary to remove all cytoplasmic materials that might obscure these features and then to mount the diatoms in a mounting medium that will provide contrast. A mounting medium with an index of refraction greater than either the glass of the microscope slide and coverslip or the silica of the diatom frustrule is used. Two techniques may be used to clean the frustrules of the cytoplasm: pyrolysis or acid digestion.

Cleaning of frustrules by acid digestion

CAUTION: Hot acid can cause severe injury to skin and eyes. Avoid breathing vapors, and provide good ventilation.

A water sample containing the diatoms is centrifuged at approximately 1,000 × g to form a pellet. The volume chosen will vary with the concentration of cells in the sample, but 20 ml may be adequate for the first attempt. Decant the supernatant water, leaving only the moist pellet. Carefully add approximately 5 ml concentrated H_2SO_4, and carefully shake to resuspend the pellet. Place the centrifuge tube with the acid and diatoms in a warm water bath (50° to 60° C) for 1 hour. Then carefully add a few drops of saturated potassium dichromate solution until the acid becomes dark brown. Remove the centrifuge tube from the water bath, and allow to cool to room temperature. Centrifuge at 1,000 × g to reform a pellet. Decant the acid, then wash and resuspend the pellet in distilled water. Centrifuge again, and rewash as many times as necessary until the decanted water is no longer noticeably acid (use pH paper or meter to check). The pH of the distilled water will probably be slightly acid before it is used to wash the diatoms. After the last resuspension, proceed with mounting.

Cleaning of frustrules by pyrolysis

Diatoms in water samples that do not have excessive organic matter other than the diatoms may be cleaned by heat oxidation. Centrifuge an appropriate volume (as described before). Wash this pellet with distilled water, which is necessary to remove dissolved salts in the water that would also be concentrated and, on evaporation, obscure the diatoms. Centrifuge again, and resuspend in distilled water. Place 1 or 2 drops of the suspension on a microscope slide, and evaporate on a warm hot plate (80° to 90° C). Do not allow to splatter. Continue adding the suspension 1 to 2 drops at a time to produce a concentrate on the slide. The organic matter in the diatoms now concentrated on the slide is oxidized by placing the slide on a hot (550° C) hot plate or in a muffle furnace at 550° C. When the residue on the slide becomes gray or very light brown, the oxidation is complete. Proceed with mounting directly on this slide as given next. NOTE: Some prefer to evaporate and incinerate the diatoms on a coverslip. However, this requires more careful temperature control during incineration because excessive temperatures will cause the coverslip to warp.

Mounting of cleaned diatom frustrules
Apparatus

Hot plate or slide warmer at a temperature of 160° C
Microscope slides and coverslips
Forceps

Reagent

Hyrax* or similar mounting medium, refractive index 1.65 +

Procedure

1. Place 1 or more drops of cleaned diatom frustrule concentrate on a coverslip, and evaporate slowly—no splattering—on a warm hot plate (ca 90° C). The number of drops will depend on concentration of the frustrules in the concentrate.

2. Place 1 drop of Hyrax on a microscope slide, and then place slide on the hot plate (ca 160° C).

*Hyrax is supplied by Custom Research & Development Inc., 8500 Mt. Vernon Rd., Auburn, Calif. 95603.

3. Allow several minutes for the solvent to evaporate from the Hyrax (the solvent keeps the Hyrax liquid at room temperature).

4. Grasp the coverslip with the deposit of diatom frustrules with forceps, and, if it has cooled, warm gently (too much heat will cause the thin glass to warp); while still warm, invert and place on the warm drop of Hyrax. NOTE: If pyrolysis on a microscope slide was used, the Hyrax is placed directly on the diatoms, then covered with a warm coverslip.

5. Immediately remove microscope slide from the hot plate, and gently apply steady pressure to the coverslip with a soft object such as a pencil eraser. If the Hyrax hardens before all the air is expelled from under the coverslip, reheat gently until just soft enough for further pressure. Remember, the thinner the slide preparation, the better for oil immersion examination.

6. Scrape excess Hyrax from slide. Label the slide.

NET PLANKTON

Net plankton may be identified and enumerated by either aliquot methods or total count methods. Total counts are usually taken of the larger forms, such as copepods and cladocerans, as well as colonial algae, such as *Volvox*. The aliquot methods are usually used for rotifers, colonial diatoms such as *Asterionella* or *Tabellaria* species, and filamentous green and blue-green algae. Copepod nauplii are also usually counted by aliquot methods. The copepod and cladoceran populations at any given time are usually restricted to three or four species of copepods and four to six species of cladocerans; usually one or two of each are dominant.

If specific identifications are to be made, it is best to identify individuals extracted from the sample before enumeration begins. Such separation and micromanipulation requires a steady hand and micromanipulators. Such micromanipulators may be made in the laboratory. Zooplankton are easily picked up by an Irwin loop, which is made by bending a very fine piece of wire about 5 cm long over a pin, then twisting the free ends until they cross over at the pin. A loop is formed when the pin is slid from the wire. This is mounted to an applicator swab stick for a handle. Several sizes of loops may be made by using various sizes of pins. Irwin loops are also available commercially. Microdissecting needles are easily made as well. Take a swab stick, and, with a small insect pin, make a shallow hole in the center of one end. Then pick up a "minuten" insect pin with a pair of fine forceps, dip one end in fast-drying glue, and insert this end into the hole made in the end of the swab stick. Several of these should be made; some can be gently dropped on their points on a hard surface to produce a hooked tip that is excellent for micrograsping.

Smaller forms by aliquot
Apparatus

> Compound microscope with mechanical stage and equipped with Whipple eyepiece disk
> Dissecting microscope
> Sedgwick-Rafter counting cell with coverslip
> Hensen-Stempel pipette or large-bore pipette with 1 to 2 ml capacity
> Petri dish, 5 to 10 cm diameter

Procedure

1. Mix preserved sample well, and withdraw aliquot by means of Hensen-Stempel pipette or by quick suction on regular large-bore pipette. Introduce this sample into Sedgwick-Rafter counting cell as illustrated in Fig. 23. When the coverslip is placed diagonally and the sample is introduced in one or both corners, a surface film of water raises the coverslip from the ground glass edges and allows it to swing about to cover the cell. Be sure no bubbles form in the corners. To avoid formation of bubbles during counting period because of warming of sample, be sure the sample is at or slightly above room tem-

Fig. 23. The Sedgwick-Rafter cell is commonly used for quantitative analysis of net plankton. One ml of sample is introduced at the corner of the cell by means of a Hensen-Stempel pipette. This pipette is designed to take a random aliquot from the plankton sample.

perature before making the slide. Allow the organisms to settle to the bottom of the chamber for several minutes.

2. Place slide on stage of microscope, and take either total, strip, or field counts as necessary to obtain required number of organisms for statistical validity, counting until average does not change or until a hundred organisms of average frequency are counted. It is quicker to count these organisms using a 10× objective with 10× oculars; however, many investigators prefer to use a 20× objective, which improves the opportunity for identification on the Sedgwick-Rafter cell. The 20× objective is

the maximum that the depth of the Sedgwick-Rafter cell will permit.

3. For many waters, three longitudinal strip counts at 100× magnification will be adequate. The upper and lower parallel sides of the Whipple eyepiece disk are used to limit the width of the counted strip. Organisms falling on the upper line are included in the count, whereas organisms falling on the lower line are excluded. When 10× objectives are used, the width of the strip counted with a Whipple disk is approximately 1 mm; however, the width should be accurately measured with a stage micrometer. If we assume a width of 1 mm, a strip length of 50 mm, and cell depth of 1 mm, each strip will have counted 50 mm^3, or $1/20$ of the volume of the cell. From these relationships, calculate the total number of each species in the 1 ml of water contained by the Sedgwick-Rafter cell.

4. These counts are for 1 ml of plankton concentrate. It is now necessary to relate these data back to the content of unconcentrated lake water. To do this, measure the volume of concentrate. Then determine the number of organisms of each species per liter of lake water as follows:

$$\text{Organisms per liter of lake water} = \frac{\text{Organisms per ml of concentrate} \times 1{,}000}{\text{Concentration factor}}$$

where

$$\text{Concentration factor} = \frac{\text{Volume of lake water filtered (ml)}}{\text{Volume of concentrate (ml)}}$$

Larger forms by total count
Procedure

1. Empty contents of concentrated preserved sample from bottle into Petri dish with grid or strips on the bottom. If the volume is too great to be contained in this dish, it may be further concentrated in the laboratory by pouring through a portion of plankton netting.

2. Allow this material to settle for a few minutes, and

then count by either total, field, or strip method. For larger forms, such as copepods and cladocerans, use a dissecting microscope with a magnification of $20\times$ or $30\times$; consequently, it is easy to count the entire sample for these large forms.

3. Obtain the number of organisms of each species per original liter of water sample by determining the volume of concentrate in the dish and dividing this into the total volume filtered (be sure to use the same units) to determine a concentration factor.

Reporting plankton results

The results of plankton counts are usually reported as number of organisms per liter of original water sample. However, because of the great range of sizes occurring in the plankton community, such reports may misrepresent the relative importance of a species. Obviously, as in other fields of ecology, one would prefer to use biomass or even energy units for most purposes. This is impractical, although single zooplankton have been weighed on ultrasensitive balances. It is possible to provide a reasonable estimate of the biomass of each major component of the community by computing the volumes of the organisms. Since the specific gravity of living plankton tissue is essentially 1, a direct volume-weight relation exists. To calculate volume, one must measure the principal dimensions of the organisms with the Whipple disk micrometer or other micrometer. After measuring several "randomly selected" individuals, an average for each taxon may be determined. The volume of the "average" organism of each taxon is then computed, assuming that it approximates some standard geometric solid such as a cylinder (Volume $= \pi\, r^2 h$), a cone (Volume $= \pi\, r^2 h/3$), or a sphere (Volume $= 4\pi\, r^3/3$).

PLANKTON CHLOROPHYLL

Following is a general survey method for estimating the concentration of phytoplankton chlorophyll pigments in a volume of water. The value of such data is uncertain, since many variables are present. However, among the many attempts to use such data are characterization of community age and structure (Odum, McConnell, and Abbott, 1958), quantification of phytoplankton standing crop (Small, 1961), and photosynthetic rates (Ryther and Yentsch, 1957).

The primary degradation of chlorophyll is the loss of the magnesium atom from the tetrapyrole portion of the molecule. The resulting molecule, a phaeophytin, will also absorb light at the chlorophyll wavelengths. Presence of this product will result in overestimation of functional chlorophyll. In healthy, growing phytoplankton communities, this error is probably negligible. However, if large quantities of degradation products are suspected, such as at the climax of a "bloom," the quantity of phaeophytin should be determined and subtracted from the calculated chlorophyll concentrations.

Apparatus

Spectronic 20 with red phototube and filter
1-inch matched test tubes
1-inch regular test tubes
Rubber stoppers or corks to fit 1-inch regular test tubes (must be
    soaked in dilute acetone for 24 hours to leach out any color)
Filteration apparatus
0.8 μm, 47 mm filters or GF/C glass fiber filters, 47 mm
Centrifuge with stoppered *glass* centrifuge tubes

Reagent

90% alkalized acetone. Using reagent grade acetone and distilled water, prepare a 90% solution. Add about one spoonful of powdered magnesium carbonate per liter of acetone solution.

Procedure

1. Measure depth of photic zone with submarine light meter between 10 AM and 3 PM when sun is not occluded by clouds.

2. Take water samples at several random locations over the lake and at several depths down to limit of the

photic zone. Usually 1-liter samples will be adequate, but 2 liters or more may be required for oligotrophic waters.

3. Filter a portion of a well-shaken sample through a 0.8 μm, 47 mm membrane filter or glass fiber filter. Because membrane filters become clogged suddenly, in contrast to the gradual decrease in filtering rate that occurs with glass fiber filters, pour small portions of lake water into the receiving funnel, waiting for each portion to be nearly filtered before adding more water. When the filtering rate begins to decrease perceptibly, the filter is about to clog completely. At this point allow suction to damp dry the filter, and then *carefully* remove receiving funnel. Do not decrease suction until edge of filter has been lifted to vent suction flask. At this time suction may be turned off, and the filter carefully removed. Do not touch plankton on filter; hold by edges only.

4. Roll up or fold filter with plankton inside. Place filter in bottom of dry, 1-inch test tube. (If desired, the filter with plankton may be frozen in a stoppered tube for several days' storage at this point).

5. Measure the volume of water filtered. Save remainder of unfiltered sample.

6a. *If membrane filter was used,* add 20 ml 90% alkalized acetone to sample tube containing filter and plankton. Stopper tube with an acetone-leached stopper. Shake until filter is dissolved. If acetone is not a distinct green or olive color, repeat steps 3 to 5, and place additional filter in the tube with the first one until solution is visibly green. Solubility of membrane filters is about one 47 mm filter per 5 ml acetone. Therefore, four is the maximum usable number of filters.

6b. *If glass fiber filter was used,* transfer the glass fiber filter to a tissue grinder, and grind in 3 to 4 ml acetone. After grinding, transfer to a centrifuge tube washing grinder with 2 to 3 ml fresh acetone, and, combining with first portion in centrifuge tube, stopper for extraction.

7. Allow extraction to proceed in a dark refrigerator (4° to 5° C). An overnight extraction is preferable.

8a. *If membrane filter was used* and any turbidity is apparent in extract, centrifuge extract in stoppered tubes.

8b. *If glass fiber filter was used,* centrifuge and decant the clear green supernatant fluid into a 10 ml graduated cylinder, and make up to 10 ml with 90% alkalized acetone.

9. Use a blank of acetone containing the same number of filters used for the sample, and measure absorbance (abs.) of sample and blank at 750 and 663 nm, using the matched test tubes. Additional measurements at 665, 645, and 630 nm are required if the trichromatic equations for a narrow band pass spectrophotometer are to be used (opposite page). In acetone extracts the absorbance of this turbidity blank at 750 nm may be appreciable.

Calculations

For grass-green extracts:

Chlorophyll a concentration (mg per liter of extract) =
$$(\text{abs. at } 663 - \text{abs. at } 750) \times 7.5^*$$

Chlorophyll *a* is the chief chlorophyll measured at 663 nm; however, the band width of the B & L Spectronic 20 probably also measures some chlorophyll *b* when it is present. Consequently, estimates of chlorophyll *a* by this method will usually be higher than those obtained by the trichromatic equations given later.

Chlorophyll a concentration of original lake water (mg/m³) =
$$\frac{\text{Chlorophyll in extract (mg/liter)} \times \text{Extract volume (ml)}}{\text{Filtered volume (liters)}}$$

If a narrow band pass (<5 nm) spectrophotometer is available, the trichromatic equations from Strickland and Parsons' handbook (1968) are preferable. These not only improve accuracy of chlorophyll *a* determination but also allow estimations of chlorophylls *b* and *c*.

*7.5 is used for 1-inch test tubes; 13.4 is used for 1 cm test tubes.

Procedure and calculations for use with a narrow band pass spectrophotometer

Prepare 20 ml extract as before, but measure absorbance at 750, 665, 645, and 630 nm. Subtract absorbance at 750 nm from each of the other absorbances to correct for turbidity. The following calculations are used:

Chlorophyll a concentration (C_a) in sample = 11.6 (abs. at 665 nm) −
1.31 (abs. at 645 nm) − 0.14 (abs. at 630 nm)

Chlorophyll b concentration (C_b) in sample = 20.7 (abs. at 645 nm) −
4.33 (abs. at 665 nm) − 4.42 (abs. at 630 nm)

Chlorophyll c concentration (C_c) in sample = 55.0 (abs. at 630 nm) −
4.64 (abs. at 665 nm) − 16.3 (abs. at 645 nm)

mg pigment per m³ of lake water =

$$\frac{C \times \text{Extract volume (ml)}}{\text{Liters of lake water filtered} \times \text{Path length (cm)}}$$

Correction for phaeophytin a

Acidification of chlorophyll *a* will remove the magnesium atom from the ring structure and produce the phaeophytin *a* molecule. If a pure sample of chlorophyll *a* is acidified, the absorbance at 665 nm is reduced by 40%, whereas acidification of pure phaeophytin *a* does not result in any change of absorbance. Thus a mixed sample (chlorophyll *a* + phaeophytin *a*) will have some intermediate absorbance.

Procedure

1. After determining the absorbances at 750 and 665 nm, acidify the sample in the spectrophotometer cell with 1 or more drops (depending on the volume of cell) of 1.0 N HCL.

2. Wait 1 minute and remeasure the absorbances at 750 and 665 nm.

3. Correct for turbidity by subtracting the absorbance at 750 nm from the absorbance at 665 nm for each measurement, that is, before and after acidification.

Calculate chlorophyll *a* in the presence of phaeophytin *a* as follows:

mg chlorophyll a per m³ of lake water =

$$\frac{28.9 \ (abs. \ before \ acid - abs. \ after \ acid) \times Extract \ volume \ (ml)}{Liters \ of \ lake \ water \ filtered \times Path \ length \ (cm)}$$

NOTE: The factor 28.9 is a product derived from the absorption coefficient for chlorophyll *a* at 665 nm and the change in absorption of pure chlorophyll before and after acidification. Slightly different values are used depending on wavelength (665 versus 663) and the consideration of possible products of chlorophylls *b* and *c*.

PHYTOPLANKTON PHOTOSYNTHESIS

Many different parameters can be used to determine the synthetic rate of phytoplankton. One can make direct measurements of increases in carbon, dry organic matter, plankton biomass, or dry plankton. However, the most common method measures the rate of oxygen evolved or carbon (^{14}C) uptake during photosynthesis. Remember, either of the following procedures provides valid estimates of the water's primary photosynthetic production only to the extent that the sample or samples (preferably) adequately represent the variable conditions present.

Table 11. Freshwater phytoplankton equivalents

	Carbon (1 mg)	Dry organic matter (1 mg)	Oxygen 1 ml	Oxygen 1 mg	Plankton biomass (1 mg)	Dry plankton (1 mg)
Carbon (mg)	1.0	0.54	0.53	0.375	0.11	0.50
Dry organic matter (mg)	1.85	1.0	1.0	0.69	0.20	0.92
Oxygen (ml)	1.87	1.0	1.0	0.70	0.21	0.93
Oxygen (mg)	2.67	1.44	1.44	1.0	0.29	1.34
Plankton biomass (mg)	8.9	4.8	4.7	3.3	1.0	4.45
Dry plankton (mg)	1.98	1.07	1.05	0.74	0.22	1.0

Since we know the approximate relationship of oxygen evolution to carbon reduction and carbohydrate formation, photosynthesis can be stated in various terms, equivalent to the other parameters. Table 11 indicates phytoplankton equivalents and permits the conversion of oxygen evolution data to other terms (or, indeed, interconversion between any of the various equivalent parameters of production). For example, the evolution of 1 ml of oxygen (column 3 of Table 11) is approximately equivalent to the synthesis of 0.53 mg carbon, 1.0 mg dry organic matter, 1 ml oxygen, 1.44 mg oxygen, 4.7 mg plankton biomass, and 1.05 mg dry plankton.

CALCULATION OF INTEGRAL PHOTOSYNTHESIS

It is often desirable to express photosynthetic production (as well as other parameters) on the basis of that occurring beneath an area of lake surface, often 1 m^2. This is done by plotting volume-based measurements from several depths against the depth from which they were sampled (Fig. 24). These data points are then connected by a straight line (trapezoidal method) or by a smooth curve. The smooth curve is usually preferred. A typical photosynthesis-versus-depth curve for a eutrophic lake in which there is surface inhibition would be significantly underestimated using a straight-line plot. The area under the curve is then determined by integration. Since the axes of this plot have values of a g C/m^3/day and meters, the area of a square is the product of these; consequently, g C/m^2/day. Integration then may be done with a planimeter or counting of the known-area squares (the smaller each square, the more accurate the integration); if a straight line is used, calculate the area beneath each pair of adjacent depths as the area of a trapezoid (rectangle plus triangle).

CALCULATION OF PHOTOSYNTHETIC EFFICIENCY

If solar energy flux data are available for the time of incubation, the efficiency of photosynthetic conversion

Fig. 24. Example of computing integral values. Here the area below a curve of photosynthesis versus depth is integrated to determine integral photosynthesis under a square meter of lake surface. The trapezoidal integration is the simplest in that straight lines are used to connect data points, shown here as dashed lines. The area of each trapezoid (rectangle plus associated triangle), 0 to 1 m, 1 to 2 m, etc., is determined and summed for the total area. Although adequate for many situations, trapezoidal integration may not give as valid a determination as some other method of connecting the data points that may better approximate the true conditions present. In this example, the smooth curve *(heavy solid line)* probably better represents the distribution of photosynthetic activity in this productive lake with maximum photosynthesis at some subsurface depth. Using the smooth curve, the integral photosynthesis is 676 mg C/m²/day, whereas trapezoidal integration is less at 659 mg C/m²/day. In this example, each square as shown has an areal value of (50 mg C/m³/day) (0.5 m) = 25 mg C/m²/day. Thus integral photosynthesis = 27 squares × 25 mg C/m²/day = 675 mg C/m²/day (note that any square one half or more below the line is included). The value given above, obtained by planimetry, is only 1 mg different.

of the energy can be calculated. These efficiencies are usually less than 2% for freshwater communities. Energy flux data are available from the National Oceanic and Atmospheric Administration (NOAA) for many stations across the United States, one of which may be near the water being studied. However, because of probable variation in cloud conditions, it is usually preferable for measurements to be taken on site with a portable recording radiometer. Data are usually reported as calories, ergs, or langleys, which may be interconverted (Appendix 1).

Calculation

$$\text{Photosynthetic efficiency } (\%) = \frac{\text{g C/m}^2\text{/day} \times 2 \times 5,500}{\text{Solar energy} \times 0.5} \times 100$$

where

5,500 = Approximate caloric equivalent per gram dry algal tissue
   2 = Approximate conversion of carbon to dry algal tissue
   0.5 = Conversion of total solar radiation to the portion of the spectrum that is photosynthetically active (PAR)

and

Solar energy is expressed as calories/m²/day

Oxygen method

The light- and dark-bottle method (Gaarder and Gran, 1927) of measuring the synthetic rate of phytoplankton is suitable for eutrophic waters. Precise analysis of small changes in oxygen concentration may be difficult in oligotrophic waters. This method compares the oxygen changes that occur in plankton communities contained in clear bottles with those occurring in dark bottles during a 24-hour period. In the light bottles, oxygen is evolved during photosynthesis and consumed by plant and animal respiration. In the dark bottle only respiration occurs.

Apparatus

3 glass-stoppered 125 to 300 ml reagent or BOD bottles for each measurement, one painted black (or wrapped with black

plastic tape) and wrapped in aluminum foil (provide a square of aluminum foil for final capping)

Rope or other means to suspend pairs of bottles in lake at depth of collection

Various means of attaching paired bottles to rope (snaps and rings, clamps, or wire baskets)

Water collection and analysis apparatus as described in section on dissolved oxygen

Reagents

As for dissolved oxygen.

Procedure

1. Select stations that will adequately represent the major morphological characteristics of the lake and depths that will adequately represent conditions of the water column. These depths may be at regular intervals or at percentages of surface illumination (100%, 75%, 50%, 25%, 1%).

2. Take a sample from each depth, using the best oxygen techniques. Use two light bottles and one dark bottle for each depth. It is preferable to use a large enough sampler (3 liters) to fill all three bottles.

3. Return one light and one dark bottle to their original depth for incubation. Be certain that glass stoppers are firmly in place and that foil cap completely seals area between cap and shoulder of the dark bottle. It is important to avoid even pinhole light leaks. Record the time.

4. Fix remaining bottle from each depth with Winkler oxygen reagents, and determine initial (I) oxygen concentration of the water sample by titration.

5. After 24 hours recover other bottles, and determine oxygen concentration.

Calculations

Plankton community respiration (as mg oxygen consumed per hour) $= \dfrac{I - D}{24}$

Net plankton photosynthesis (as mg oxygen per hour) $= \dfrac{L - I}{24}$

Gross plankton photosynthesis (as mg oxygen per hour) =

Net photosynthesis + Community respiration

where

I, D, and L = Oxygen concentrations as mg/L for initial bottle, dark bottle, and light bottle, respectively

These data may be converted to grams of oxygen per cubic meter by multiplying numerator and denominator by 1,000. To estimate the photosynthetic activity per square meter of lake surface, either (1) multiply cubic meter value by the depth, in meters, that it represents, and sum each of these; or (2) plot actual cubic meter values on linear graph paper against depth (m), and then integrate area beneath curve by planimeter or counting of squares.

^{14}C method

The principal advantage of measuring phytoplankton photosynthesis by the ^{14}C method rather than using the oxygen change procedures is its greater sensitivity. This permits estimation of primary production in more oligotrophic waters. Disadvantages of the ^{14}C method include (1) higher costs of instruments and supplies, (2) the need to obtain the proper license when using nonexempt quantities of the isotope, and (3) some uncertainty about the exact nature of the rates being measured. It is assumed that when short incubation periods (1 to 4 hours) are used, the rate approximates net photosynthesis. This method is also widely used to measure photosynthesis in algal bioassay studies.

Apparatus for inoculation and incubation

Incubation bottles, 125 to 300 ml glass-stoppered bottles: an equal number of dark bottles are necessary and prepared as described for oxygen method

Support for incubation bottles: as for oxygen method

Water collection bottle: inert polyvinyl chloride (PVC) type preferred but not essential for fresh waters

Total carbon bottles: BOD or similar bottles for subsequent de-
    termination of total alkalinity and pH for each depth sampled
Insulated "ice" chest painted black inside
Aluminum foil or absorbent paper to contain radioactive spills
Inoculation syringe with 4-inch cannula: a 1 ml tuberculin
    syringe with Chaney adaptor provides good replication
Rubber or disposable gloves
Containers for radioactive waste solids and liquids
Squeeze bottle with 5% HCl for decontamination of spills

Apparatus for filtration and counting

Membrane filtration apparatus and source of vacuum: 47 mm
    diameter, if counting is by liquid scintillation, or 25 mm
    diameter, if counting is by planchet
Membrane filters, 0.45 μm pore size and of correct diameter
Source of distilled water
Small foreceps for handling filters
Aluminum planchets and moderately fast-drying household
    cement (Duco), if planchet counting, or liquid scintillation
    vials, if using liquid scintillation for counting
Wax pencil or felt-tipped pen for labeling planchets or vial caps
Provision for disposal of waste liquids and solids
Liquid scintillation "cocktail" as discussed in step 9b of pro-
    cedure
Apparatus for determination of pH and total alkalinity

Reagents

Sodium carbonate—^{14}C solution, specific activity = 2 μCi/ml. For
occasional work one may purchase this as a sterile aqueous solution
in sealed ampules with sufficient volume for the desired study. For
extensive studies, it is much more economical to purchase concentrated solutions and dilute, then fill and seal ampules in your laboratory as follows: purchase $Na_2{}^{14}CO_3$ stock solution with a radioactive
concentration of 2 mCi/ml. Dilute 1 ml of stock solution to 1,000 ml
with a dilution solution (next). Fill ampules with quantity desired,
and seal. Autoclave ampules in inverted position in a metal pan filled
with a diluted methylene blue solution. Remove from autoclave,
and cool to room temperature in the pan of dye solution. Any ampules that have an imperfect seal will suck the dye solution inside.
These must be discarded. The absolute activity of the final solution
will depend on the degree of accuracy claimed by the manufacturer

of the stock solution and on your quantitive techniques. For precise work, the assumed activity must be confirmed. Such a service is provided for a fee by many commercial laboratories, or the activity may be determined in the laboratory if liquid scintillation instruments are available (Wolfe and Schelske, 1967).

Dilution solution. Dissolve 50 g NaCl in 1 liter of demineralized water; add 0.3 g Na_2CO_3 as a carrier, then add 1 to 2 pellets of NaOH to bring the pH to 9.5 to 10.0. The pH must not be less than 9.5. The NaCl allows this to be used in marine and estuarine waters, and in fresh waters it increases the density of the solution so that the solution tends to sink when injected into the sample bottle, assuring no loss when the stopper is inserted.

Procedure

1. Select locations and depths as for oxygen method. Since this procedure takes only a short time, use care in selecting time of incubation. Because of the possibility of respiration of the radioactive compounds and consequent underestimation of primary production, use the shortest time period possible that will give sufficient activity for precision in counting. A 2-hour midday period will often provide sufficient illumination. A diel series of measurements should be made to determine the relation of the selected incubation time to the full photic period. Generally, one half of the full day's photosynthesis occurs during the 4-hour period from 10 AM to 2 PM.

2. Collect water samples from desired depths. Fill one light bottle, one dark bottle, and one bottle for pH and alkalinity from the sampler. Proceed with inoculation immediately; or, if several depths are to be sampled, place bottles in blackened ice chest until all samples for the location are taken; then proceed with inoculation as soon as possible.

3. Open an ampule and fill syringe, being certain that no bubbles remain.

4. Open the light and dark bottles, and inoculate (1 ml per 125 ml bottle) the $Na_2{}^{14}CO_3$ solution at bottom of bottle, using the long cannula. Remove cannula quickly, and restopper bottle. Immediately cap black bottle with a

double thickness of foil. Inspect for tears or other possible light leaks. Begin timing incubation period.

5. Resuspend bottles at desired depth as soon as possible. A rack to maintain bottles in a horizontal position is desirable (Fig. 25). At all times, avoid exposing clear bottles to direct sunlight at the surface of the lake or while inoculating. Such exposure, especially of samples from greater depths, may cause a photoinhibition of the contained phytoplankton.

6. During incubation period, one may determine pH, total alkalinity, and temperature of sample if portable equipment is used or if a laboratory is nearby. If not, place these samples on ice for analysis on returning to the laboratory.

7. At the end of the incubation period, retrieve bottles and place in blackened ice chest. Cover them with crushed ice if field filtration is not to be done. Since samples in the iced-dark condition will not have a significant

Fig. 25. When several replicates from one depth are to be incubated for ^{14}C uptake, some type of rack to resuspend all samples horizontally at one depth with a minimum of shading is desirable. Note the foil seal over the cap of each dark bottle.

change in radioactivity for up to 3 hours, it may be preferable to return the samples to a laboratory in which better filtration and vacuum is available. If any delay is an-
ticipated, the samples should be preserved by injecting 1
ml of 50% formalin (per 125 ml bottle) immediately on
retrieval of the bottles.

8. Filter all of sample, or an aliquot thereof, depending on the amount of radioactive uptake and the presence
of silt or detritus that clogs the filter. Gentle filtration is
important to avoid cell damage and loss of tagged carbon
compounds; one-third atmosphere is probably safe for
most forms. If a vacuum gauge is not available, use
thin-walled rubber tubing rather than the usual thick-
walled vacuum tubing. Often 50 ml taken as a well-
shaken aliquot will provide sufficient radioactivity for
short but precise counts. Wash filter and apparatus with a
volume of distilled water approximately equal to the
original volume filtered. If deposition of insoluble carbonate salts on the filter is suspected (alkaline water of
high pH), remove any radioactive inorganic deposits
from the filter by rinsing with 2% HCl or fuming over
HCl vapors. Follow this procedure with a distilled water
rinse, especially if liquid scintillation counting is to be
used, since acid filters will not clear adequately in the
scintillation fluid.

9a. *Planchet counting.* With fingertip, smear a thin
film of cement on planchet, and immediately place filter
on cement. A few seconds' delay will allow a surface film
to dry, and attachment of filter may be imperfect, with
subsequent curling. Write appropriate label on bottom of
planchet. Count a known standard (to determine efficiency of counter) and the samples according to instructions for detector system used. Count to a minimum of
2,000 (approximately 5% error). A small sample size,
such as 50 ml, will usually produce a sufficiently thin
layer so that self-absorption can be ignored.

9b. *Liquid scintillation counting.* If instrumentation
is available, liquid scintillation counting offers many ad-

vantages (including a much greater efficiency) over planchet counting (Lind and Campbell, 1969). Remove filter and coil to place vertically in scintillation vial, with algal layer to the inside. Two scintillation cocktails are commonly used. One requires desiccation of filters before adding cocktail but gives higher efficiency. The second, although it has a slightly lower efficiency, does not require desiccation, which permits a more rapid completion of the procedure and recovery of data. If the former is desired, place uncapped vials in a desiccator charged with silica gel desiccant and small vial of CO_2 absorbent. When dry, add 20 ml of the cocktail: 4 g PPO (2,5-diphenyloxazole) and 0.1 g dimethyl POPOP (p-*bis*-[2-(4-methyl-5-phenyloxazoyl)] benzene) are dissolved in toluene (AR) and diluted to 1 liter. Cap vial immediately, and label the cap. A perfectly dry filter will become completely transparent in the cocktail. If you choose the method that does not require desiccation of filters, place the moist filter in the vial as before, and add 20 ml of commercially available water-tolerant cocktails (e.g., Aquasol) or the following cocktail: dissolve 4 g PPO, 0.1 g dimethyl POPOP, 80 g crystalized naphthalene, and 50 ml *p*-dioxane in toluene (AR), and dilute to 1 liter. Cap the vial, label, and count as before. A disadvantage of this method is the slow solution of the filter in dioxane, which cancels out the advantage of the vertical filter.

10. Wipe fingerprints from vial before placing it in counter. Count a standard (to determine efficiency) and samples according to instruction of manufacturer. Rarely, if algal concentrations are great or if sample contained large amounts of silt or debris, quenching may occur. Correct for this by preparing a quench curve made up from constant amounts of a ^{14}C standard solution, such as toluene-^{14}C, which is added to vials containing filters through which progressively greater amounts of the quench material–containing water have been filtered. Instructions for quench correction vary with different instruments and must be made according to the manufacturer's instructions.

Calculations

Rates of phytoplankton photosynthesis may be calculated according to the following:

$$P = P_l - P_d$$

and

$$P_l \text{ or } P_d = \frac{r}{R} \times C \times f$$

where

P = Phytoplankton photosynthesis in mg carbon per cubic meter

P_l or P_d = Carbon uptake per cubic meter in light and dark bottle, respectively

r = Uptake of radioactive carbon in counts per minute = Counts per minute for filtered volume $\times \dfrac{\text{Volume of bottle}}{\text{Volume filtered}}$

R = ^{14}C inoculated in counts per minute = $(2.22 \times 10^6) \times$ Microcuries added $\times$ Efficiency of counter

C = Inorganic carbon (^{12}C) available in mg per cubic meter = Total alkalinity $\times$ Conversion factor from Table 12 $\times$ 1,000

f = Correction of slower uptake of ^{14}C as compared with ^{12}C = 1.06

For example, assume that a 125 ml light bottle is inoculated with 2 μCi ^{14}C; 50 ml are filtered and give a count of 4,500 per minute at an efficiency of 85%. The pH is 7.5, and the total alkalinity is 100 mg/L at 20° C.

$$P_l = \frac{4,500 \times \dfrac{125}{50}}{2.22 \times 10^6\,(2)\,(0.85)} \times (100 \times 0.26 \times 1,000) \times 1.06 =$$

82.90 mg of carbon uptake in the light per cubic meter per incubation time

The calculation for the corresponding dark bottle uptake proceeds the same. The difference between P_l and P_d approximates net phytoplankton photosynthesis.

COMMUNITY METABOLISM

The metabolic activity of any system consists of two kinds of energy transformation: anabolic (synthetic) and

Table 12. Factors for conversion of total alkalinity to milligrams of carbon per liter*

pH	Temperature (° C)					
	0	**5**	**10**	**15**	**20**	**25**
6.0	1.15	1.03	0.93	0.87	0.82	0.78
6.1	0.96	0.87	0.77	0.73	0.70	0.67
6.2	0.82	0.74	0.68	0.64	0.60	0.58
6.3	0.69	0.64	0.59	0.56	0.53	0.51
6.4	0.60	0.56	0.52	0.49	0.47	0.45
6.5	0.53	0.49	0.46	0.44	0.42	0.41
6.6	0.47	0.44	0.41	0.40	0.38	0.37
6.7	0.42	0.40	0.38	0.37	0.35	0.35
6.8	0.38	0.37	0.35	0.34	0.33	0.32
6.9	0.35	0.34	0.33	0.32	0.31	0.31
7.0	0.33	0.32	0.31	0.30	0.30	0.29
7.1	0.31	0.30	0.29	0.29	0.29	0.28
7.2	0.30	0.29	0.28	0.28	0.28	0.27
7.3	0.29	0.28	0.27	0.27	0.27	0.27
7.4	0.28	0.27	0.27	0.26	0.26	0.26
7.5	0.27	0.26	0.26	0.26	0.26	0.26
7.6	0.27	0.26	0.26	0.25	0.25	0.25
7.7	0.26	0.26	0.25	0.25	0.25	0.25
7.8	0.25	0.25	0.25	0.25	0.25	0.25
7.9	0.25	0.25	0.25	0.25	0.25	0.25
8.0	0.25	0.25	0.25	0.25	0.24	0.24
8.1	0.25	0.25	0.24	0.24	0.24	0.24
8.2	0.24	0.24	0.24	0.24	0.24	0.24
8.3	0.24	0.24	0.24	0.24	0.24	0.24
8.4	0.24	0.24	0.24	0.24	0.24	0.24
8.5	0.24	0.24	0.24	0.24	0.24	0.24
8.6	0.24	0.24	0.24	0.24	0.24	0.24
8.7	0.24	0.24	0.24	0.24	0.24	0.24
8.8	0.24	0.24	0.24	0.24	0.23	0.23
8.9	0.24	0.24	0.23	0.23	0.23	0.23

*From Saunders, G. W., F. B. Trama, and R. W. Bachmann. 1962. Evaluation of a modified C-14 technique for shipboard estimation of photosynthesis in large lakes. Great Lakes Res. Div. Publ. No. 8. University of Michigan, Ann Arbor.

catabolic (respiratory). These are dynamic processes. The metabolic process is both a result and a cause of change. Odum (1956) developed the concepts of autotrophy and heterotrophy in communities. Autotrophy applies to communities in which the long-term synthesis exceeds respiration; heterotrophy applies to communities in which the long-term respiration exceeds synthesis. These internal processes of synthesis and respiration may be supplemented, to varying degrees, by gain or loss from external sources.

The metabolism of the community is, in reality, the sum of the metabolic activity of each organism in that community. A metabolic study without reference to the individual constituents is relatively easily accomplished and provides much basic information about the state of the community. Carbon dioxide is taken from the water, and oxygen is added by the process of photosynthesis. Respiration continues throughout the light and dark periods, adding carbon dioxide and using oxygen. As a result, a daily pattern in the concentration of either dissolved gas is observed. A change in the concentration of either gas is used to estimate the rate of formation or the respiration of organic matter.

The diel oxygen method is actually a derivative of the light- and dark-bottle method of Gaarder and Gran (1927). In community studies the entire lake serves as the bottle, with the daylight period analogous to the light bottle and the night hours analogous to the dark bottle. A simplified method (McConnell, 1962) is suitable for small bodies of water with minimum surface agitation.

The areal hypolimnetic oxygen deficit is also a community metabolism estimate that measures the rate of respiration (primarily decomposition) in a lake's hypolimnion, which is related to the autotrophic synthesis of organic matter in the epilimnion.

Simplified diel oxygen method
Apparatus and reagents

As required for dissolved oxygen determination (pp. 73 and 74)

Procedure

Sampling depths and sites are chosen as described for phytoplankton photosynthesis. The procedure involves three samples: two taken on successive evenings and one at the interim morning. These correspond to the high and low concentrations of oxygen found in the lake water. Oxygen concentrations rise during the day but reach a rather steady level in the late afternoon. The afternoon sampling should occur during this plateau. A sample taken an hour before sunset best satisfies this requirement. Oxygen concentrations decline during the night. However, upward changes in concentration occur rapidly as soon as light becomes available. Therefore, the time of the dawn sampling is more critical. This time is usually within an hour before sunrise.

Oxygen concentrations of the water are determined for the selected sites and depths at each of these three periods.

Calculations

The three values obtained for each depth at each station are plotted as milligrams per liter against time. A solid line is drawn from the evening value to the following sunrise value and then extrapolated as a broken line for the 24-hour period to represent total decline in oxygen content because of respiration. A solid line is drawn from the dawn value to the second evening value and extended to a point equivalent to 24 hours after the first value. The difference at the 24-hour point between the top and bottom line is gross community photosynthesis as milligrams per liter of oxygen. The difference between the first and the succeeding 24-hour value of the respiration slope is total community respiration as milligrams per liter of oxygen. An example of community metabolism calculation is shown in Fig. 26.

Data as milligrams per cubic meter from each depth may be plotted against depth (in meters) on linear graph paper and integrated by planimetry or by counting

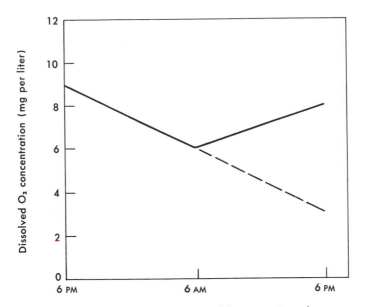

Fig. 26. Example of diel oxygen data used for estimation of community metabolism. At 6 PM of the first day the dissolved oxygen concentration at one station and depth was 9.0 mg/L. At 6 AM of the following day the dissolved oxygen concentration had declined to 6.0 mg/L. At 6 PM of the second day, the oxygen concentration increased to 8 mg/L. The broken line is an extrapolation of the night respiration slope. Total community respiration = 9 mg/L − 3 mg/L = 6 mg/L. Gross community photosynthesis = 8 mg/L − 3 mg/L = 5 mg/L. The ratio of photosynthesis to respiration (P/R) = 0.83.

squares to calculate the rate of gross photosynthesis or respiration beneath a square meter of lake surface for the depth sampled.

Hypolimnetic areal relative oxygen deficit

Another way to consider the metabolism of the entire lake community is to compute the rate at which oxygen is consumed in the hypolimnion. The use of the hypolimnetic oxygen deficit was proposed by Strom (1931) as an indirect measure of the production. It has subsequently been used in attempts to describe the trophic status of several lakes. The technique is subject to several variables, including temperature, concentration of oxygen, and morphometry, although the latter is minimized by use of areal deficits.

The fundamental assumption is that the oxidation of organic matter (oxygen consumption) in the hypolimnion is relatable to the production of organic matter in the overlying epilimnion. This assumes no input to the lake of externally produced (allochthonous) organic matter, which is, of course, incorrect to varying degrees relative to the surface area of the lake, the nature of the watershed, and other variables unique to the system being studied. It also assumes that all organic matter sinking into the hypolimnion is oxidized each year and that there is no residue carried over from one year to the next.

Apparatus and reagents

As required for determination of dissolved oxygen (pp. 73 and 74) and temperature (p. 30)

Procedure

1. Measure the dissolved oxygen concentration and temperature at frequent depths through the metalimnion and hypolimnion. Begin these measurements at the onset of thermal stratification.

2. Repeat these determinations at weekly intervals (extremely eutrophic lakes may require more frequent measurements). In large lakes or lakes with unusual basin configurations, multiple sites will be required to adequately represent the hypolimnetic conditions.

3. From morphometric data (pp. 13 to 15) determine the area of the surface of the hypolimnion and the volume of the hypolimnion or strata of the hypolimnion from which oxygen samples were taken. For example, if oxygen concentrations are determined at 1 m intervals, vol-

umes of each of these meters will be needed. Usually the area and volumes near the upper portion of the hypolimnion will need to be recalculated because the surface of the hypolimnion typically descends through time. It is important to use a consistent method of determining the upper boundary of the hypolimnion, since small variations here can cause signficant differences in areal oxygen content. One method defines the upper limit of the hypolimnion as the point derived from depth-temperature plots at which a line extrapolated upward from the linear portion of the curve well within the hypolimnion intercepts a line extrapolated from the portion of the curve in the metalimnion with the maximum rate of temperature change.

4. Determine the oxygen content of the hypolimnion

Fig. 27. Example of a plot for the calculation of the relative areal hypolimnetic oxygen deficit. $y = 2.01 - 0.066x$; relative areal hypolimnetic oxygen deficit = 0.066 mg $O_2/cm^2/day$.

by multiplying the measured concentration of each stratum by the volume of that stratum and summing the figures for all strata in the hypolimnion.

5. Convert these volume-based data to areal basis by dividing the content of the hypolimnion by the surface area of the hypolimnion in square centimeters.

6. Plot the content as milligrams oxygen per square centimeter against time, using the date of the first sample as day one (Fig. 27). Some form of a linear decrease ($y = a - bx$) is expected. The daily areal relative oxygen deficit is then "b" in this equation with units of milligrams oxygen per square centimeter per day. NOTE: This relation will depart from linearity when the oxygen concentration becomes sufficiently low so as to restrict the metabolism of the oxidizing biota.

4 ☐ Benthos

Benthos are organisms that inhabit the bottom substrate of lakes, ponds, and streams. These organisms may artificially be divided into two major groups: macrobenthos and microbenthos. Macrobenthos consist of organisms retained by a no. 30 U.S. Series sieve. The following procedures are concerned only with macrobenthos that inhabit the mud, sand, gravel, or litter substrate of the lake or stream bottom. Procedures for organisms found on solid bodies projecting above the bottom, such as logs, growing vegetation, and large rocks (aufwuchs), are given on pp. 169 to 172.

Benthic organisms play several important roles in the aquatic community: they are involved in the mineralization and recycling of organic matter produced in the open water above or brought in from external sources, and they are important second and third links in the trophic sequence of aquatic communities. Many benthic insect larval forms are major food sources for small fishes.

Quantitative benthic investigations usually determine the number and kinds of organisms present for the estimation of standing crop and production and serve as indicators of water quality. Benthic organisms are especially useful in pollution studies, and indices of their diversity are widely used for estimation of the degree of pollution. The procedures given in this chapter include methods for the collection, concentration, separation, identification, and enumeration of these organisms.

COLLECTION OF BENTHIC SAMPLES
Lentic

A variety of lentic bottom dredges (grabs) are available. Each was originally designed for one specific sampling problem. Several common collecting instruments are the Ekman dredge, the Petersen dredge, and the Ponar dredge (Fig. 28). The Ekman dredge is the easiest to use in that it is light and relatively easy to "set." However, its use is limited to soft mud, silt, or finely divided sand bottoms. Because of its relatively small sampling area, one must take many replicate samples. For sampling where the bottom material is compacted or consists of gravel, rock, or organic litter substrate, the Petersen or the Ponar dredge is preferred. These dredges are quite heavy and

Fig. 28. Commonly used benthos dredges in the open position.

therefore normally require a winch. The weight will vary between 15 and 30 kg, depending on the number of weights applied. Small obstacles to the closing of these dredges are crushed by the jaws, whereas the same materials would block the operation of the Ekman dredge.

A precaution to be noted with any of these types of dredges is the possible disturbance of the surface film of the bottom substrate as the dredge is lowered. Although the Ekman dredge employs a hinged top that allows the column of water to pass through, there is always some displacement of water. With the Petersen dredge, this displacement is considerable. It produces an outward blastlike current on the finely divided substrate just before the dredge settles. To overcome this problem, the dredge must be lowered gently through the last 0.5 m.

Although the closing of the Petersen or Ponar dredge takes a considerably deeper bite than does the Ekman, each samples only the first few centimeters of substrate. It has been demonstrated that in many waters the bottom fauna is distributed for depths considerably below the bite of these dredges. In shallow waters (up to 3m), it is possible to fit an Ekman dredge with a solid rod mechanism rather than the rope and messenger used in deeper waters. With this modification, one can lower the Ekman dredge onto the bottom and then force it deeper into the sediments as it is being tripped. However, in any investigation it is probable that a certain portion of the deep-burrowing benthos is missed by any of these commonly used sampling techniques.

When these dredges are raised, it is important that they not be lifted clear of the water and allowed to drain, since small organisms will be lost through the small seams from which the water runs. Therefore, the best procedures require a lifting of the dredge from the surface of the water while slipping a bucket beneath the dredge to catch these materials.

Apparatus

Dredge, line, and messenger
Two or more metal 10- to 12 liter pails
Metal laundry tubs
Large-diameter no. 30 screen
Wide-mouthed bottle (500 to 1,000 ml)

Reagent

Supply of neutral formalin

Procedure

1. Select sampling sites representative of the area to be sampled, and determine number of replications

A

Fig. 29. A, Contents of a Ponar sample are emptied into a large tub.

necessary. Be sure to record number of replicate samples taken.

2. Set dredge trip mechanism after securing other end of line to some fixture of the boat. These dredges are especially heavy when loaded, and precautions should be taken to avoid dropping and loss.

3. Lower dredge to sampling site, lowering slowly through the final 0.5 m. If depth is unknown, lower to bottom and then raise, move 1 or 2 m laterally, and re-lower gently. Trip dredge by dropping messenger for the Ekman or by allowing the line to slacken for the Petersen or Ponar.

4. Lift the filled dredge to the surface with a smooth,

B

Fig. 29, cont'd. B, Water is added, and large particles are broken up by hand.

Continued.

Fig. 29, cont'd. C, This slurry is then poured through a benthos screen.

even motion to avoid jarring out contents, and swing into a pail as mentioned previously.

5. Dump contents of dredge and pail into a washtub (Fig. 29, A). If the boat is large, the following procedures may be done in the boat; if not, return filled washtub to shore, and do following procedures there. Repeat the sampling procedure until all the replicates have been taken.

6. Reach into the tub, and break up all compacted particles (Fig. 29, B). These are especially common in areas of clay soils. Continue working the material in the tub until it is of a fine homogeneous consistency.

7. Pour the contents of the tub (use a pail to dip when

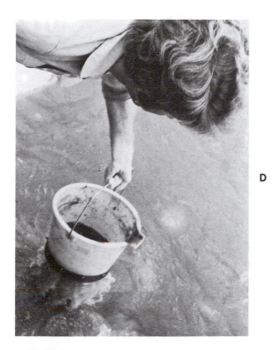

Fig. 29, cont'd. D, Sieving of the screen contents is accomplished by sloshing in and out of the water with a twisting or swirling motion.

the tub is full) into the no. 30 screen, holding the screen over the edge of the boat (Fig. 29, *C*). Continue this step until all the contents of the tub have been poured through the screen and the tub and any pails used have been rinsed several times with lake water.

8. The material remaining on the screen is your collected sample. Use a sloshing, twisting, and swirling motion as you thrust the screen up and down into the water (Fig. 29, *D*), but do not let the water run over the top of the screen. This will wash smaller materials through the sieve and at the same time will tend to concentrate the materials at one edge if you work holding the screen on a slant.

9. When all materials are concentrated in one area, scrape these into a wide-mouthed bottle, wash screen again, and place any additional materials into the bottle, picking out the last few remains with fine forceps if necessary.

10. Preserve this material with adequate neutral formalin to produce a 5% to 10% solution (one part formalin to nine parts water, ooze, and organisms).

Lotic

Sampling from running waters differs in both the nature of the organisms collected and the means used for the collection. The distinction between benthos and aufwuchs is poorly defined in running waters. Because of the scouring action of the current, soft sediments are

Fig. 30. The Surber stream sampler is used for quantitative benthos analysis of shallow streams and rivers. Placed parallel with the current, the bottom substrate is stirred within the area of the frame. Dislodged organisms are washed into the net. (Courtesy Wildlife Supply Company, Saginaw, Mich. 48602.)

rarely found. Organisms of running waters are either heavy bodied or have special means of attachment. Finally, because of the lack of fine sediments, the distribution of organisms with depth in the sediment is usually greater in running waters than in standing waters.

The most commonly used sampling device for running waters is the Surber stream bottom sampler (Fig. 30). This device has a major drawback in that its use is restricted to waters of less than 30 cm in depth and of slow to moderate velocity. Because quantitative sampling devices are lacking for other types of flowing water, the only method suitable for such situations is an indirect measurement of the benthic fauna by measurement of drift using a series of drift nets.

Surber sampler for shallow streams

Apparatus

Surber stream bottom sampler
Garden trowel
Stiff bristle brush (fingernail type)
Wide-mouthed bottle (500 to 1,000 ml)

Reagent

Neutral formalin

Procedure

1. Select sampling site representative of area desired and with no depth greater than height of sampling frame. The velocity of flow must not be so great as to cause a "pressure head" of water to flow around mouth of net.

2. Wade into water from downstream, and place net with the mouth facing upstream in an undisturbed area. Be sure there is no disturbance of the bottom substrate upstream from the net, or organisms will be dislodged and washed into the net.

3. Lower square foot frame on substrate, and hold it in place. Pick up larger rocks or bits of substrate, and,

while holding them in mouth of net, brush them free of all organisms, allowing current to carry them into net. Discard these rocks outside the frame.

4. When all larger bits have been brushed, use garden trowel to stir up all the substrate within the square foot frame. Be careful that the stirring motion is toward center of frame so that organisms that are dislodged will be carried into mouth of net and not around it. Attempt to stir this area thoroughly and to a uniform depth.

5. Depending on the number of replicates desired, move the sampler to other sites, and repeat steps 1 through 4.

6. After all replicates have been taken, empty contents of net into a wide-mouthed jar. The net must be turned "wrong side out," and organisms that have attached to the fabric picked from net and placed into jar.

7. Add sufficient water to cover the substrate and organisms in the bottle and then sufficient formalin to make a 5% to 10% solution. Label jar, indicating number of replicates, and return to the laboratory.

Drift nets for shallow or deep streams

Drift nets collect organisms being carried downstream (drift) by the currents. They are usually used in a series placed perpendicular to the flow. If the total cross-sectional area of all nets is known and the cross-sectional area of the stream measured (pp. 18 and 19), one can estimate the percent of stream flow being sampled. Sufficient nets should be used to adequately represent stream conditions, that is, slow, swift, shallow, deep, and so forth. The nets may be left in place for 24 hours to sample daily drift, since there are usually great diel differences. However, for most streams, this time would greatly overload both the nets and the time of the investigator for sorting and counting. It is usually preferable to set the nets for 1- to 2-hour intervals several times throughout the day.

Apparatus

Set of drift nets with stakes and lines for anchoring
Wide-mouthed bottles (500 to 1,000 ml)

Reagent

Neutral formalin

Procedure

1. Set drift nets either vertically or horizontally to representatively sample the stream. They may be tied to stakes driven into the stream bottom or to tightly stretched lines fastened from bank to bank across the stream. The top of shallow nets should be just below the surface to avoid floating detritus; the bottom of deep nets should be just above the stream bottom to prevent organisms from crawling in, since such are not "drift."

2. After an appropriate time, retrieve the nets and empty contents into the wide-mouthed jars. Preserve with enough neutral formalin to make a 5% solution. Label the jars with the collection information.

LABORATORY PROCESSING OF BENTHIC SAMPLES

The collected samples contain a mixture of mud, rocks, sand, debris, and the desired organisms. The next step is to remove the organisms from the unwanted material and separate them into similar taxonomic groupings for final identification and enumeration. A variety of laboratory techniques, each specifically suited for a particular type of sample, is used by different investigators. Techniques described next are general and should be modified where such a modification would lend improvement. It is important for the student to keep in mind that the procedures used must be nonselective so that the final list will be truly representative of the habitat sampled.

Apparatus

Graded series of soil sieves
Shallow white enamel pans
Fine-tipped, gentle-action forceps
Large glass funnel (20 to 25 cm in diameter) with large drain
    opening (3 to 5 cm) fitted with flexible rubber tubing and
    hose clamps
Dissecting microscope with good illumination
Series of small vials or baby food jars

Reagents

Table sugar (several pounds)
4% neutral formalin solution

Procedure

If the collected sample contains much extraneous material, such as gravel or organic debris (which is usually the case with Surber samples), proceed as follows. If the sample has been well screened in the field, omit the laboratory screening procedures and go directly to step 3.

1. Place the graded series of soil sieves into a sink, and empty sample or part of sample into top screen. Allow a gentle stream of water to flow through screen to thoroughly wash and sort materials into different screens. Be careful that bottom screens do not plug, causing a backflushing and loss or organisms over the top.

2. Remove upper screen of very coarse materials, and pick through it for very large organisms. As organisms are separated, place them into vials or baby food jars containing 4% formalin according to their approximate taxonomic groupings.

3. Organisms from screens may be separated from the rubble by several methods. Each is best suited for particular types and frequency of organisms and the nature of the rubble. Three methods follow.

3a. Empty each screen with its contained rubble and organisms into a shallow enamel pan, add enough water to disperse the materials, and using forceps, pipette, or small metal scooping screen, take off organisms from rubble, stirring frequently to ensure that all are loosened and separated from the debris. Check material from finer screens under a dissecting microscope to be sure that no small organisms are being missed.

3b. Place materials from screens in white enamel pan, and flood with sugar solution. This solution should be made up of table sugar and tap water and should have a specific gravity of 1.12 (approximately 300 g sugar per liter). Make this deep enough to cover the debris by at least 1 inch. Stir debris, allow currents to stop, and soft-bodied forms will float to the surface, where they may be scooped off. Repeated stirrings are necessary to allow materials to separate. Complete this process within 15 minutes, or organisms will begin to sink. Heavy-bodied organisms, such as fingernail clams, stone-cased caddis flies, and oligochaetes, must be picked off the bottom. Pour material remaining in tray back through a screen of equal or smaller size, and collect the sugar solution, which should be saved and used many times. Flood debris with tap water, and allow it to soak for 15 minutes; then repeat sugar flotation procedure after tap water has been poured off. This second floating serves as a check on your separation. Always make the final inspection with a dissecting microscope. When sugar solution becomes strongly discolored so that it is not possible to see through it to bottom of pan, it may be clarified by running through a pad of gauze or coarse filter paper. This sugar solution may be reused (until it is no longer possible to clear it) by making only necessary adjustments to the specific gravity by addition of more sugar.

3c. The following procedure is also a sugar separation technique. Mount large funnel in ring stand (Fig. 31). The rubber tubing is connected to the bottom and closed with pinch clamp. Fill this funnel approximately half full of the 1.12–specific gravity sugar solution as in 3b. Empty contents of one screen into this funnel, and stir. Allow contents to settle, and most debris will sink

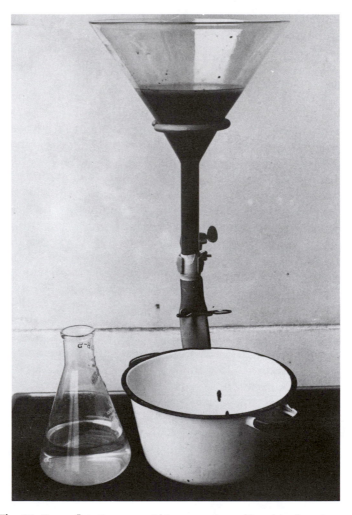

Fig. 31. Sugar flotation may aid in separation of benthos from heavy debris. Here a sample is added to a funnel containing a sugar solution. This solution is stirred well and then allowed to settle. Debris and an occasional heavy-bodied organism are drained into one container, while the supernatant fluid with floating organisms are separated into a second container. This separation is repeated on the first settlings.

into neck of funnel. Squeeze the soft rubber tubing in a pumping motion to set up currents, which will tend to dislodge organisms trapped in the debris. Allow contents to separate for a few seconds, then skim off organisms from top of funnel. Drain funnel through tubing into two vessels in the same manner one would use a separatory funnel, placing heavy debris in one and supernatant into a second. Be sure to wash down sides of funnel with sugar solution into the second container. The supernatant can then be passed through a fine screen to collect any small organisms that were not skimmed from the top. After this has been done, repeat the procedure on same material that has been collected in the first container.

4. In every case, after organisms have been separated from the debris, they should be washed free of any sugar solution, placed in approximate taxonomic groupings (if possible), and preserved in 4% neutral formalin.

5. Identification of most bottom fauna requires the use of a dissecting microscope with good illumination. Pennak's *Fresh-water Invertebrates of the United States* is a good starting point for identification. Edmondson's *Fresh-water Biology* should also be available.

6. After organisms have been identified, count them by taxonomic group, and report as numbers per area. Determine the area sampled by the dredge used, and multiply this by the number of replicates taken to get the total area sampled. Data should be standardized and reported as number of organisms per square meter.

7. Further processing may be desirable in that some data are better expressed as biomass. In this case it is necessary to wash the organisms in tap water to remove the formalin and blot dry with absorbent tissue for a wet "live" weight, or place in shallow dishes and dry in a laboratory oven at 103° to 105° C, and cool in a desiccator before weighing for dry weight. Note that the preservative (concentration or type) may affect the relation of laboratory weight to live weight.

AUFWUCHS (PERIPHYTON)

The aufwuchs may be considered a special case of the benthos. The aufwuchs make up any community that develops on the surface of any submerged substrate. Many investigators use this term interchangeably with the term periphyton. The aufwuchs community developing on hard or permanent, usually rock substrates can be distinguished as epilithic aufwuchs; the community developing on organic, usually plant substrates are epiphytic.

By definition, then, this community will have a highly variable composition; although it may include such large members as snails and insect larvae, investigators are more typically concerned with the microscopic members. Protozoa, rotifers, and nematodes are especially common animals, whereas the diatoms and blue-green algae are the common plants. Colonial, stalked, and gelatinous forms are common among both plants and animals. Many species in the typical aufwuchs community are delicate, soft-bodied organisms. Therefore, special care must be taken in handling and preserving them. In fact, it is often preferable to examine this community alive and undisturbed.

Although relatively difficult to investigate, this community provides an excellent biological detector of water quality. Because of their sessile nature, the aufwuchs cannot move to escape onslaughts of pollution; thus these organisms are constant, round-the-clock detectors of pollution.

Study of the aufwuchs may be undertaken from natural substrates or by colonization of artificial substrates. Whereas artificial substrates have many practical advantages, especially for quantitative studies, it is probable that a "natural" community never develops on an artificial substrate in any reasonable period of time.

Once collected, identification and counting of aufwuchs proceeds much the same as for plankton. The Sedgwick-Rafter cell may be used on well-mixed samples for larger forms; sedimentation techniques and gentle membrane filtration may be used for smaller forms.

Collection methods
NATURAL SUBSTRATES

Qualitative sampling may be done by gently scraping the material from any submerged surface into a small collection jar or vial. If examination cannot be accomplished within a day, preservation in 5% formalin is necessary. Quantification of natural substrates is difficult because of their irregular surface areas. Obviously, the surface area of a regular object, such as the cylindrical stem of a bullrush, may be quantified if the entire surface area is scraped into the collecting jar. Other, more irregular surfaces may be quantified as follows.

Apparatus

Isolating cylinder (large-diameter cork borer ideal; however, any type of cylinder that can be beveled to sharp edge satisfactory)
Stiff fingernail brush
Scraping device, such as a spatula or knife
Vials or small baby food jars

Reagent

Supply of 5% formalin

Procedure

1. Select the natural substrate—wood, stone, and so forth—that appears to have been in its present position for some time. Remove from water, but keep surface wet.

2. Using the cork borer or other isolating cylinder, press down a "randomly" selected portion of the substrate, and, while holding borer firmly in place, brush away all aufwuchs around the cylinder for an area of at least 2 inches. Wash this area off with water to be certain that all surrounding organisms have been removed. At

this point, you have isolated an island of aufwuchs of known area.

3. Gently scrape this isolated island into the small collection vial with a small amount of water. If possible, return vial to the laboratory while keeping it cool, and proceed with identification and counting of the living material. If identification and counting must be postponed, preserve organisms in 5% formalin.

ARTIFICIAL SUBSTRATES

Many types of artificial substrates, including concrete blocks, bricks, and building tiles, have been used for aufwuchs studies. These materials have the advantage of having regular, known area surfaces that can be entirely scraped clean for quantitative determinations. Another type of artificial substrate that allows the study of the aufwuchs community without damage by scraping is the glass microscope slide.

Apparatus

Support for microscope slides. Any type of device that will rigidly hold a series of separated microscope slides in a vertical position, and, if in a flowing system, with the surfaces parallel to current's flow, will suffice. This is accomplished by using a wooden microscope slide file box from which both top and bottom are removed, leaving only ends and sides with their cut divisions for separating slides. Top and bottom are covered with rectangular pieces of ¹/₄- to ¹/₂-inch (0.6 to 1.2 cm) mesh hardware cloth. Once slides have been inserted in box, hardware cloth is held in position by heavy rubber bands or cord.

Large-mouthed, 2- to 4-liter jars of sufficient size to contain slide box rack.

Ropes, anchors, and floats necessary to suspend slide box artificial rack.

Reagent

Supply of formalin sufficient to produce a 4% to 5% solution when added to the large-mouthed container full of water

Procedure

1. Number microscope slides, using a diamond-tipped scribe, permanent ink, or paint. Place slides in rack, and firmly bind so that they will not be dislodged during a long colonization period.

2. Suspend artificial substrate rack in the lake or river in such a way that slides are held in a vertical position and, if there is current, with flat surfaces parallel to direction of current. A series of these slides through various depths may be desired, since distinct depth distributional patterns occur among the aufwuchs, with autotrophic components dominating near the surface and heterotrophic components dominating in the deeper waters.

3. Allow artificial substrates to colonize for a minimum of 2 weeks. Colonization time may need to be longer, depending on the season of the year and water temperature, as well as the nature of the study.

4. After colonization, gently remove entire sampler from water, and place in large-mouthed jar to be returned to the laboratory for live examination if possible, or preserve in 4% formalin if necessary. A preserved slide has the advantage in that one may work directly with oil immersion on the preserved slide without a coverslip.

5. In the laboratory, scrape one surface of microscope slide clean with a razor blade. This surface will be the "bottom" and will be placed on the microscope stage. While keeping upper surface wet, scan entire microscope slide with both a dissecting microscope and a compound microscope.

6. Take quantitative counts, using a Whipple eyepiece disk, counting all organisms contained in random fields (by using a mechanical stage, you may take strip counts). It is also possible to scrape off all the aufwuchs components on the surface of the slide except for a known area and then take a total count of the larger organisms for the known area.

7. Biomass may be determined by scraping entire surface of microscope slide with a razor blade and col-

lecting the aufwuchs in an evaporating dish, in which the organisms are dried and weighed, using the techniques described in the section on residue on pp. 95 to 99.

DIVERSITY OF BENTHOS

Diversity indices relate the number of kinds of organisms to the total number of organisms and in some cases to the number of individuals of each kind. Such indices may be applied to any biotic community but have had widest application with the benthos. A statement of species diversity has become widely used as an indicator of water quality: low diversity often indicates water of low quality. Undisturbed natural communities are assumed to have a high diversity, that is, a relatively large number of species, with no species having disproportionately large numbers of individuals.

A number of calculations, each with specific applications, have been developed to express diversity. One of the simplest is diversity $(D) = m/N^{1/2}$ where m is the total number of species, and N is the total number of individuals. Although this describes the species richness of the community, it omits an important consideration: the evenness with which N is distributed among m, or the equitability component. Any diversity index that emphasizes richness is more sample size dependent, an undesirable feature. An index based primarily on the equitability of the common species is preferred. The Shannon index (Shannon and Weaver, 1949) satisfies this condition and is calculated as diversity $(H) = -\Sigma P_i(\ln P_i)$. P_i is the importance probability of each species and is equal to n_i/N, where n_i is the number of individuals (or biomass) of each species, and N is the total number (or biomass) of species.

Any of these methods requires a taxonomic knowledge of the organisms. One type of diversity index, the sequential comparison index, does not have this requirement, thus permitting the most inexperienced limnologist to use it (Cairns and others, 1968).

Sequential comparison index
Procedure

1. Combine all organisms from a given collection. Use some method to arrange the organisms randomly in a straight line (Fig. 32). For example, you might add a small amount of water and empty the organisms into a rectangular white enamel pan. Wash down to one side with a stream of water from a squeeze bottle, then blot off excess water.

2. Begin at one end of the row of organisms. Pick them out one at a time, and decide if this organism is different from the preceding organism. If it is not, go on to the next one; if it is different, record this as the start of a new "run." Therefore a new run occurs when any organism is different from the preceding one. At the end of the row, count up the number of runs and the total number of organisms.

Fig. 32. A portion of a randomly arranged row of stream benthos for calculation of the sequential comparison index. This segment of 11 organisms contains 8 obviously different runs. A difficult judgment must be made to determine if the small mayfly (fifth from left) is different enough from the preceding larger mayfly to count it as a separate run.

3. Calculate the diversity index (DI):

$$DI = \frac{\text{Number of runs}}{\text{Number of organisms}}$$

4. If the sample size is excessively large (greater than 250 organisms), stop at the end of the first 50 organisms and calculate DI, that is, number of runs/50. Then continue for the next 50 organisms and recalculate: number of runs/100. On graph paper make a plot of calculated DIs on the ordinate against number of specimens on the abscissa. Continue calculating and plotting at the end of each 50 organisms until the curve obtained becomes obviously asymptotic. At this point there is no advantage in continuing. A statistical analysis of such data indicates that when sufficient organisms are present in the sample as described in this step, a rerandomization and repeat of the procedure is sufficient to produce a $\overline{DI}$ with a 95% confidence interval equal to $\pm 10\%$ of calculated $\overline{DI}$.

PRODUCTION OF BENTHOS

Methods for the measurement of secondary production of benthos are not as generally applicable as those for primary production. Methods for measurement of secondary production must be quite specific for each group of organisms. Differences in length of life span, presence or absence of juvenile instars, carried or demersal eggs, size-(age-)specific or nonspecific mortality. population size, and turnover rates all must be considered. Consequently, it is difficult to present a single method. Following is a method of applicability to organisms with a single generation per year (univoltine), for which the turnover rate, birthrate, or death rate is unknown. It is best applied to a single taxon but may be applied to the collective univoltine benthic community.

Procedure

1. Collect organisms quantitatively from a known area, and separate by taxon. Prepare for dry-weight determinations as described on p. 167, step 7.

Table 13. Example of tabulation for calculation of benthic production

Size group (mg)	Collections (no./area)					N	$\bar{n}$
	1	2	3	4	5		
0.0 to 0.49	150	130	100	70	20	470	94
0.5 to 0.99	60	40	25	5	2	132	26
1.0 to 1.49	50	40	20	10	2	122	24
1.5 to 1.99	10	10	9	3	0	32	6

2. Weigh dried organisms, and tabulate numbers into weight groups. Considering smallest and largest organisms, determine magnitude of weight groupings so that organisms can be separated into 4 to 10 groups (Table 13).

3. Repeat this procedure for a series of equally spaced dates for the same habitat. Sampling effort should be equal for each date. If calculation of annual production is desired, this series might be twelve monthly samplings. NOTE: If a major increase, that is, greater than one attributable to sampling variation, in the smallest-sized group between successive dates occurs, recruitment has occurred, indicating a polyvoltine species. In such an event, the second group must be treated separately for the balance of the period of study, and the production of each age group of the species summed for total production.

4. Tabulate as in the example shown in Table 13 for only four size groups and five collections.

Calculations

Calculate production according to:

$$P_B = i \, \Sigma(\bar{n}_j - \bar{n}_{j-1}) \frac{\overline{w}_j + \overline{w}_{j-1}}{2}$$

where

P_B = Biomass production per m² per time
i = Number of collections

n_j = Number of organisms of size group j per m²
$\overline{n}_j$ = n_j/number of collections
$\overline{w}_j$ = Mean weight of size group j

therefore

$$P_B = 5[(94 - 26)\left(\frac{0.25 + 0.75}{2}\right) + (26 - 24)\left(\frac{0.75 + 1.25}{2}\right) +$$

$$(24 - 6)\left(\frac{1.25 + 1.75}{2}\right) + 6(2.0)] = 375 \text{ mg per m}^2 \text{ per 5 time units}$$

P/B ratio

Following is the production-to-biomass (P/B) ratio for the preceding example:

$$P/B = \frac{\text{Annual production}}{\text{Mean annual biomass}} = \frac{375 \text{ mg per m}^2}{83.5 \text{ mg per m}^2} = 4.49$$

It is often practical to compute the P/B ratio. This relates the annual production to some biomass parameter. Often mean annual biomass is used, although some prefer to use maximum spring biomass for univoltine species. The P/B ratio then allows for the estimation of production in that specific community in future years from biomass data only.

MACROPHYTES

The term *macrophyte* refers to all the large macroscopic plants of lakes, streams, and wetlands. Although primarily vascular plants that are usually, but not always, rooted, the term also includes large algae such as *Chara* and *Nitella* species, and mosses such as *Fontinalis* species. Techniques for many of these are the same as those used for terrestrial vascular plants. The principal differences are in the collection of deeply submerged plants such as *Potamogeton, Elodea,* and *Chara* species and in the drying of the especially succulent soft tissues. These plants of the littoral community are often encrusted with dense growths of epiphytic aufwuchs, which must be considered in any study of biomass or production.

Qualitative studies

Apparatus

Plant grapple. This may be purchased or made by folding a 50 cm
piece of heavy-gauge wire in half to make a very narrow
U-shaped piece. Three or four of such pieces are then forced
through a 15 cm piece of 3/4- to 1-inch (2 to 3 cm) pipe and
the U-shaped portions exposed for 3 to 5 cm. These should
have a very tight fit in the pipe. The free ends are then bent
back on the pipe to about a 45-degree angle with the pipe at
approximately equal intervals around the pipe. The U-shaped
portions are then tied together with a wire or line to which a
sampling "cast" line may be attached.

Vasculum, bucket, or several plastic bags, into which the plants
may be placed without drying out.

Boat, if necessary.

Sonar depth sounder ("fish-finder"), if available and necessary.

Procedure

1. Plant beds are located visually, or they may be
found in deep water by using the depth sounder. A broad
or multiple "echo" indicates plant beds.

2a. In shallow water, in which hand collection is possible, collect the entire plant including root, although
cutting may be necessary, since many aquatics have extensive rhizomal systems. Attempt to obtain material in
flowering and fruiting condition. Look for heterophylly,
that is, variation in leaf form, and, if present, collect
plants representing all forms.

2b. In deep water, plants are collected by use of the
grapple. Attach the grapple to a "cast" line, and throw the
line out from the boat, allow it to sink to the bottom, and
retrieve through the plant bed with a horizontal-to-
oblique tow. This technique typically yields only shoot
material, although some roots may be pulled from the
sediment. Attempt to obtain a variety of material, as described in 2a. Of course, submerged macrophytes may be
obtained by diving, especially with the aid of snorkel or
scuba equipment.

3. Wash as much attached material from the roots and

shoots as possible, then place along with a waterproof label in the vasculum, bucket, or bags, and return to the laboratory.

4. In the laboratory, many identifications may be made from wet materials, with no further processing necessary. However, if materials are to be retained in a collection or mailed away for identification, dry herbarium mounts are usually made, although some macrophytes, especially macroalgae, mosses, and very small forms, are best preserved in liquid preservatives such as 4% neutral formalin or FAA (10 ml formalin, 50 ml 95% ethanol, 2 ml glacial acetic acid, and 40 ml distilled water).

5. Because many aquatic macrophytes lack supportive tissues, special procedures are required to get them spread out sufficiently for useful herbarium mounts. To accomplish this, place a sheet of paper (newspaper or herbarium paper) in the bottom of a shallow rectangular pan. Add about 3 to 5 cm water to the pan, keeping the paper on the bottom. Then place the well-washed plant specimen on the water in the pan, and guide it to floating in a somewhat natural, spread-out position. Next, gently lift the paper from the pan, allowing the plant to settle on the paper (a little practice will produce good results).

6. Place the plant on the paper, along with its original label, between two sheets of newspaper, which are then placed between two herbarium blotters and spacers. This is then placed in a press and dried in a herbarium dryer. Note that the spongy tissues of many aquatic plants will require longer drying than terrestrial plants.

7. When completely dry, remove from the press and papers. Mount the plant on herbarium paper with strips of tape or glue, and label with a permanent label affixed to the herbarium sheet.

Quantitative studies

Apparatus

Quadrat grid: metal or weighted wooden square of 0.25 to 0.5 m area

Large knife, grass clippers, and trowel to cut off or dig up plants
Plastic bags
Snorkel or scuba equipment, necessary for deep-water collec-
    tions
Boat, if necessary
Sonar depth sounder ("fish-finder"), if available and necessary

Procedure

1. Locate plant beds visually or with sonar.

2. For quantitative studies care must be taken to collect both roots and shoots for total plant tissue, or to collect only shoot tissue cut off at the surface of the sediments. For deep water, this requires experienced divers
(in a team, never alone). "Randomly" place the grid over
the plants, and work it down to the bottom. Often, care is
necessary to prevent shoots from crossing under the grid.
Only plants rooted within the grid will be harvested.

3. With the grid in place, harvest all plant tissues
above the sediments by clipping or cutting with the
knife. Plants should be immediately placed in a plastic
bag. However, most will float to the surface so that lost
material may be recovered. If total plant tissue is desired,
use the trowel to dig up entire root systems after
rhizomes have been cut along the perimeter of the grid
with the large knife. Place a collection label in the bag.
Return to the laboratory.

4. Empty the contents of the bag into a fine mesh
sieve, such as is used in benthos studies, and carefully
wash to remove sediments and epiphytic growths.

5. The material from one grid is then blotted damp
dry for estimation of "live weights" or oven dried at
103° C to a constant weight. The results are usually
expressed as grams dry weight per square meter.

Useful
conversion factors

To convert	Into	Multiply by
acre	hectare	0.4047
acres	square feet	43,560.0
acres	square meters	4,047.0
acre-feet	cubic feet	43,560.0
acre-feet	cubic meters	1,234.0
acre-feet	gallons	3.259×10^5
angstrom units	microns	1×10^{-4}
atmospheres	inches of mercury (0° C)	29.92
atmospheres	kilograms/square centimeter	1.033
atmospheres	pounds/square inch	14.70
British thermal units	ergs	1.055×10^{10}
British thermal units	gram calories	252.0
Celsius	Fahrenheit	$(1.8C) + 32$
centimeters	feet	3.281×10^{-2}
centimeters	inches	0.3937
centimeters/second	feet/second	0.03281
cubic centimeters	cubic feet	3.531×10^{-5}
cubic centimeters	cubic inches	0.06102
cubic centimeters	cubic meters	1×10^{-6}

To convert	Into	Multiply by
cubic centimeters	gallons (U.S. liquid)	2.642×10^{-4}
cubic feet	pounds of water	62.4
cubic feet	cubic inches	1,728.0
cubic feet	cubic meters	0.02832
cubic feet	gallons (U.S. liquid)	7.48052
cubic feet	liters	28.32
cubic feet/second	million gallons/day	0.646317
cubic feet/second	gallons/minute	15.9
cubic inches	cubic centimeters	16.39
cubic inches	cubic meters	1.639×10^{-5}
cubic inches	liters	0.01639
cubic meters	cubic feet	35.31
cubic meters	cubic yards	1.308
cubic meters	gallons (U.S. liquid)	264.2
cubic meters	liters	1,000.0
einsteins/square centimeter/second	quanta/square centimeter/second	6.02×10^{23}
ergs	gram calories	0.2389×10^{-7}
ergs	kilocalories	2.389×10^{-11}
ergs/second	kilocalories/minute	1.433×10^{-9}
ergs/square centimeter/second	gram calories/ square centimeter/second	2.387×10^{-8}
Fahrenheit	Celsius	$^5/_9(F - 32)$
fathom	meter	1.828804
fathoms	feet	6.0
feet	centimeters	30.48
feet	kilometers	3.048×10^{-4}
feet	meters	0.3048
footcandle	lumen/square meter	10.764
footcandles	lux	10.764
gallons	cubic centimeters	3,785.0
gallons	cubic feet	0.1337
gallons	cubic meters	3.785×10^{-3}
gallons (Br. imp.)	gallons (U.S.)	1.20095
gallons	liters	3.785
gallons of water	pounds of water	8.3453
gallons/minute	cubic feet/second	2.228×10^{-3}

To convert	Into	Multiply by
gallons/minute	liters/second	0.06308
grains	grams	0.0648
grains/gallon	parts/million	17.118
grams	grains	15.43
grams	ounces (avoirdupois)	0.03527
grams	ounces (troy)	0.03215
grams	pounds	2.205×10^{-3}
grams/centimeter	pounds/inch	5.600×10^{-3}
grams/liter	grains/gallon	58.417
grams/liter	parts/million	1,000.0
grams/square centimeter	pounds/square foot	2.0481
grams/square meter	pounds/acre	8.992
gram calories	British thermal units	3.9683×10^{-3}
gram calories	ergs	4.1868×10^{7}
gram calories	watt hours	1.163×10^{-3}
gram calories/square centimeter/second	ergs/square centimeter/second	4.19×10^{7}
hectares	acres	2.471
hectares	square feet	1.076×10^{5}
inches	centimeters	2.540
inches	meters	2.540×10^{-2}
joules	ergs	1×10^{7}
joules	kilocalories	2.389×10^{-4}
kilograms	pounds	2.205
kilograms/cubic meter	pounds/cubic foot	0.06243
kilograms/hectare	pounds/acre	0.8922
kilograms/meter	pounds/foot	0.6720
kilograms/square meter	pounds/square foot	0.2048
kilometers	centimeters	1×10^{5}
kilometers	feet	3,281.0
kilometers	miles	0.6214
kilometers/hour	feet/second	0.9113
kilowatts	kilocalories/minute	14.34
kilowatt-hours	gram calories	859.85×10^{3}
knots	kilometers/hour	1.8532
knots	statute miles/hour	1.151
liters	cubic feet	0.03531

To convert	Into	Multiply by
liters	cubic meters	1,000.0
liters	gallons (U.S. liquid)	0.2642
liters	quarts (U.S. liquid)	1.057
liters/minute	cubic feet/second	5.886×10^{-4}
$\log_{10}$	$\log_e$	2.303
$\log_e$	$\log_{10}$	0.4343
lumens/square foot	footcandles	1.0
lux	footcandles	0.0929
meters	feet	3.281
meters	inches	39.37
meters	miles (statute)	6.214×10^{-4}
meters	yards	1.094
meters/minute	feet/second	0.05468
meters/second	miles/hour	2.237
microns	meters	1×10^{-6}
miles (statute)	kilometers	1.609
miles (statute)	meters	1,609.0
miles (statute)	miles (nautical)	0.8684
miligrams/liter	parts/million	1.0
millimeters	feet	3.281×10^{-3}
millimeters	inches	0.03937
millimicrons	meters	1×10^{-9}
million gallons/day	cubic feet/second	1.54723
ounces	grams	28.349572
ounces	pounds	0.02957
ounces (U.S. fluid)	liters	29.6×10^{-3}
parts/million	grains/U.S. gallon	0.0584
pounds	grains	7,000.0
pounds	grams	453.5924
pounds	kilograms	0.4536
pounds	ounces	16.0
pounds of water	cubic feet	0.01602
pounds of water	cubic inches	27.68
pounds of water	gallons	0.1198
pounds/acre	grams/square meter	0.112
pounds/cubic foot	kilograms/cubic meter	16.02
quanta/square centimeter/second	einsteins/square centimeter/second	1.66×10^{-24}
quarts (liquid)	cubic centimeters	946.4

To convert	Into	Multiply by
quarts (liquid)	cubic feet	0.03342
quarts (liquid)	cubic meters	9.464×10^{-4}
quarts (liquid)	liters	0.9463
square centimeters	square feet	1.076×10^{-3}
square centimeters	square inches	0.1550
square centimeters	square meters	0.0001
square feet	acres	2.296×10^{-5}
square feet	square centimeters	929.0
square feet	square meters	0.09290
square inches	square centimeters	6.452
square kilometers	acres	247.1
square kilometers	hectares	100.0
square kilometers	square meters	1×10^6
square kilometers	square feet	10.76×10^6
square kilometers	square miles	0.3861
square meters	acres	2.471×10^{-4}
square meters	hectares	1×10^{-4}
square meters	square centimeters	1×10^4
square meters	square feet	10.76
square meters	square miles	3.861×10^{-7}
square meters	square yards	1.196
square miles	square kilometers	2.590
square miles	square meters	2.590×10^6
square yards	square meters	0.8361
tons (metric)	kilograms	1,000.0
tons (metric)	pounds	2,205.0
watts	ergs/second	107.0
watts	joules/second	1.0
watts	kilocalories/minute	0.01433
watts/square centimeter	langleys/minute	14.3
watt-hours	ergs	3.60×10^{10}
watt-hours	gram calories	859.85
yards	centimeters	91.44
yards	kilometers	9.144×10^{-4}
yards	meters	0.9144

APPENDIX TWO **Sources of limnological apparatus and supplies**

This partial list implies no endorsement of product or service. A more complete listing is *Sources of Limnological and Oceanographic Apparatus and Supplies* (special publication no. 1, third revision) available from the American Society of Limnology and Oceanography.

Reagents

Harleco
   60th and Woodland Ave.
   Philadelphia, Pa. 19143

Test kits and reagents

Hach Chemical Co.
   Box 907
   Ames, Iowa 50010
Hellige, Inc.
   877 Stewart Ave.
   Garden City, N.Y. 11533

Radioisotopes

Amersham-Serale Corp.
   2636 S. Clearbrook Dr.
   Arlington Heights, Ill. 60005

New England Nuclear
  575 Albany St.
  Boston, Mass. 02118

General sampling and field equipment

Wildco
  2200 S. Hamilton St.
  Saginaw, Mich. 48602
Kahl Scientific Instrument Co.
  P. O. Box 1166
  El Cajon, Calif. 92022
Hydro-Bios Apparatebau
  Wismarerstrasse 14
  23 Kiel, Germany
Hydro Products
  4930 Naples Place
  San Diego, Calif. 92110
Foerst Mechanical Specialties
  2407 N. St. Louis Ave.
  Chicago, Ill. 60647

Submarine photometers, thermometers, and other supplies

Montedoro-Whitney Corp.
  P. O. Box 1401
  San Luis Obispo, Calif. 93401
Kahl Scientific Instrument Co.
  P. O. Box 1166
  El Cajon, Calif. 92022

Membrane filters and supplies

Millipore Corp.
  Bedford, Mass. 01730

Nets and screens

General Biological Supply (Turtox)
  8200 S. Hoyne Ave.
  Chicago, Ill. 60620

References

American Public Health Association. 1976. Standard methods for the examination of water and wastewater. American Public Health Association, Inc. New York. 1,193 pp.

Brown, E., M. W. Skougstad, and M. J. Fishman. 1970. Methods for collection and analysis of water samples for dissolved minerals and gases. U.S. Geological Survey, U.S. Department of the Interior, Washington, D.C. 166 pp.

Cairns, J., D. W. Albaugh, F. Busey, and M. D. Chanay. 1968. The sequential comparison index—a simplified method for nonbiologists to estimate relative differences in stream pollution studies. J. Water Poll. Control Fed. **40**:1607-1613.

Correll, D. S., and H. B. Correll. 1972. Aquatic and wetland plants of southwestern United States. Environmental Protection Agency, Washington, D.C. 1,777 pp.

Eddy, S., and H. C. Hodson, 1961. Taxonomic keys to the common animals of north central states. Burgess Publishing Co., Minneapolis. 162 pp.

Edmondson, W. T., ed. 1959. Fresh-water biology, 2nd ed. John Wiley & Sons, Inc., New York. 1,248 pp.

Edmondson, W. T., and G. G. Winberg. 1971. A manual on methods for the assessment of secondary productivity in fresh waters. International Biological Program Handbook No. 17. Blackwell Scientific Publications Ltd., Oxford, England. 358 pp.

Ellis, M. M., B. A. Westfall, and M. B. Ellis. 1948. Determination of water quality. Research Report No. 9. Fish and Wildlife Service, U.S. Department of the Interior, Washington, D.C. 122 pp.

Environmental Protection Agency. 1974. Methods for chemical analysis of water and wastes. Environmental Protection Agency, Water Quality Office, Analytical Quality Control Laboratory, Cincinnati.

Eyles, D. E., and J. L. Robertson. 1963. A guide and key to the aquatic plants of the southeastern United States. Public Health Bulletin No. 86. Bureau of Sport Fisheries and Wildlife, Washington, D.C. 151 pp.

Fassett, N. C. 1960. A manual of aquatic plants. University of Wisconsin Press, Madison. 405 pp.

Gaarder, T., and H. H. Gran. 1927. Investigations of the production of plankton in the Oslo Fjord. Rapports et Procès—Verbaux des Réunions Conseil Permanent International pour l'Exploration de la Mer **42**:1-48.

Golterman, H. L., and R. S. Clymo. 1969. Methods for chemical analysis of fresh waters. International Biological Program Handbook No. 8. Blackwell Scientific Publications Ltd., Oxford, England. 188 pp.

Haney, J. F., and D. J. Hall. 1973. Sugar-coated *Daphnia:* a preservation technique for Cladocera. Limnol. Oceanogr. **18**:331-333.

Huntsberger, D. V., and P. Billingsley. 1977. Elements of statistical inference, 4th ed. Allyn & Bacon, Inc., Boston. 385 pp.

Hutchinson, G. E. 1957. A treatise on limnology, vol. 1. John Wiley & Sons, Inc., New York. 1,015 pp.

Hynes, H. B. N. 1970. The ecology of running waters. University of Toronto Press, Toronto. 555 pp.

Lind, O. T., and R. S. Campbell. 1969. Comments on the use of liquid scintillation for routine determination of ^{14}C activity in production studies. Limnol. Oceanogr. **14**:287-289.

Maciolek, J. 1962. Limnological organic analysis by quantitative dichromate oxidation. Research Report No. 60. Bureau of Sport Fisheries and Wildlife, Washington, D.C. 61 pp.

McConnell, W. J. 1962. Productivity relations in carboy microcosms. Limnol. Oceanogr. **7**:335-343.

Menzel, D. W., and N. Corwin. 1965. The measurement of total phosphorus in seawater based on the liberation of organically bound fractions by persulfate oxidation. Limnol. Oceanogr. **10**:280-282.

Moore, E. W. 1939. Graphic determination of carbon dioxide and the three forms of alkalinity. J. Am. Water Works Assoc. **31**:51.

Murphy, J., and J. Riley. 1962. A modified single solution method for the determination of phosphate in natural waters. Anal. Chim. Acta. **27**:31-36.

Needham, J. G., and P. R. Needham. 1962. A guide to the study of fresh-water biology, 5th ed. Holden-Day, Inc., San Francisco. 108 pp.

Odum, H. T. 1956. Primary production in flowing waters. Limnol. Oceanogr. **1**:102.

Odum, H. T., W. McConnell, and W. Abbott. 1958. The chlorophyll "A" of communities. Publ. Inst. Mar. Sci. Texas **5**:56-96.

Patrick, R., and C. W. Reimer. 1966. The diatoms of the United States, vol. 1. Monographs of the Academy of Natural Sciences, Philadelphia. 688 pp.

Pennak, R. W. 1978. Fresh-water invertebrates of the United States, 2nd ed. John Wiley & Sons, Inc., New York. 803 pp.

Poole, H. H., and W. R. G. Atkins. 1926. On the penetration of light into sea water. J. Mar. Biol. Assoc. U.K. **14**:177-198.

Prescott, G. W. 1962. Algae of the western Great Lakes area. William C. Brown Co., Publishers, Dubuque, Iowa. 977 pp.

Prescott, G. W. 1970. How to know the fresh water algae, 2nd ed. William C. Brown Co., Publishers, Dubuque, Iowa. 348 pp.

Rainwater, F. H., and L. L. Thatcher. 1960. Methods for collection and analysis of water samples. U.S. Geological Water Supply Paper 1454. U.S. Government Printing Office, Washington, D.C. 301 pp.

Ricker, W. E. 1968. Methods for assessment of fish production in fresh waters. International Biological Program Handbook No. 3. Blackwell Scientific Publications Ltd., Oxford, England. 326 pp.

Ryther, J. H., and C. S. Yentsch. 1957. The estimation of phytoplankton production in the ocean from chlorophyll and light data. Limnol. Oceanogr. **2**:281-286.

Saunders, G. W., F. B. Trama, and R. W. Bachmann. 1962. Evaluation of a modified C-14 technique for shipboard estimation of photosynthesis in large lakes. Great Lakes Res. Div. Publ. No. 8. University of Michigan, Ann Arbor. 61 pp.

Schindler, D. W. 1969. Two useful devices for vertical plankton and water sampling. J. Fish. Res. Board Can. **26**:1948-1955.

Schwoerbel, J. 1970. Methods of hydrobiology (fresh-water biology). Pergamon Press Ltd., Oxford, England. 200 pp.

Shannon, C. E., and W. Weaver. 1949. The mathematical theory of communication. University of Illinois Press, Urbana, Ill. 117 pp.

Small, L. F. 1961. An optical density method of measuring phytoplankton standing crop. Iowa State J. Sci. **35**:343.

Smith, G. M. 1950. The fresh water algae of the United States, 2nd ed. McGraw-Hill Book Co., New York. 719 pp.

Sokal, R. R., and F. J. Rohlf. 1969. Biometry: the principles and practice of statistics in biological research. W. H. Freeman & Co. Publishers, San Francisco. 776 pp.

Stainton, M. P., M. J. Capel, and F. A. J. Armstrong. 1974. The chemical analysis of fresh water. Spec. Publ. 25. Fish. Res. Board Can. 125 pp. + suppl.

Steele, R. G. D., and J. H. Torrie. 1960. Principles and procedures of statistics. McGraw-Hill Book Co., New York. 481 pp.

Stephens, K. 1963. Determination of low phosphate concentrations in lake and marine waters. Limnol. Oceanogr. **8**:361-362.

Strickland, J. D. H., and T. R. Parsons. 1968. A practical handbook of sea water analysis. Fisheries Research Board of Canada, Ottawa. 311 pp.

Strom, K. M. 1931. Feforvatin. A physiographic and biological study of a mountain lake. Arch. Hydrobiol. **22**:491-536.

Vollenweider, R. A. 1969. A manual on methods for measuring primary production in Aquatic environments. International Biological Program Handbook No. 12. Blackwell Scientific Publications Ltd., Oxford, England. 213 pp.

Welch, P. S. 1948. Limnological methods. McGraw-Hill Book Co., New York. 381 pp.

Wolfe, D. A., and C. L. Schelske. 1967. Liquid scintillation and geiger counting efficiencies for carbon-14 incorporated by marine phytoplankton in productivity measurements. J. Cons. Perm. Inter. Explor. Mer. 31:31-47.

Index

□ Page numbers in italics refer to illustrations; those followed by t indicate tables.

191